Technician Class

FCC License Preparation for Element 2 Technician Class Theory

by
Gordon West
WB6NOA

Seventh Edition

Master Publishing, Inc.

Also by Gordon West, WB6NOA

General Class
FCC License Preparation for
Element 3 General Class Theory
Also available as an audio book on CD
and with W5YI Ham Study software

Extra Class
FCC License Preparation for
Element 4 Extra Class Theory
Also available as an audio book on CD
and with W5YI Ham Study software

GROL + RADAR
General Radiotelephone Operator License
Plus Ship Radar Endorsement
with Pete Trotter, KB9SMG and Eric Nichols, KL7AJ
FCC Commercial Radio License Preparation for
Element 1, Element 3, and Element 8 Question Pools
Also available with
W5YI Comm Study Software

For more information visit
www.W5YI.org
or call 800-669-9594

This book was developed and published by:
Master Publishing, Inc.
Niles, Illinois

Editing by:
Pete Trotter, KB9SMG

Photograph Credit:
All photographs that do not have a source identification are either from the author, or Master Publishing, Inc. originals.
Cover photo and additional photos by Julian Frost, N3JF.

Cartoons by:
Carson Haring, AC0BU

CD mastering by:
Mike Koegel, N6PTO, Studio 49

Thanks to the following for their assistance with this book: Suzy West, N6GLF; Ed Collins, N8NUY; AMSAT and TAPR; John Johnston, W3BE; California Rescue ARES Net members; Question Pool Committee members; WA6TWF Super System members; and Don Arnold, W6GPS.

Printing by:
Arby Graphic Service
Niles, Illinois

6019 W. Howard Street
Niles, Illinois 60714
847/763-0916 (voice)
847/763-0918 (fax)
E-Mail: MasterPubl@aol.com

Visit Master Publising
on the Internet at:
www.masterpublishing.com

Seventh Edition
5 4 3 2 1

Table of Contents

QUESTION POOL NOMENCLATURE

The latest nomenclature changes and question pool numbering system recommended by the Volunteer Examiner Coordinator's question pool committee (QPC) for question pools have been incorporated in this book. The Technician Class (Element 2) question pool has been rewritten at the middle-school reading level. This question pool is valid from July 1, 2010 until June 30, 2014.

FCC RULES, REGULATIONS AND POLICIES

The NCVEC QPC releases revised question pools on a regular cycle, and deletions as necessary. The FCC releases changes to FCC rules, regulations and policies as they are implemented. This book includes the most recent information released by the FCC at the time this copy was printed.

Preface

Welcome to the fabulous, fun hobby of Amateur Radio! It has never been easier to enter the amateur radio service than now!

Changes to Amateur Radio regulations announced by the Federal Communications Commission on December 30, 1999, that became effective April 15, 2000, have made it easier than ever to obtain your entry-level Technician Class ham radio license — and to move up through all the license classes to earn the top Extra Class license.

And when I say, "It has never been easier...," I really mean it! In years past, to achieve Technician Class level, you would have studied over 900 Q & As. Most recently, the study pool was 510 Q & As. Now, beginning July 1, 2010, we are down to middle-school-reading-level, 394 questions in the pool, and just 35 on the test. Absolutely no knowledge of Morse code is required for this entry-level license, and math skills are limited to simple division and multiplication.

In just one exam session, you can satisfy all of the requirements to earn your Technician Class amateur radio license. Within days of passing your written exam (you need to answer 26 of the 35 questions correctly — that's just 74%) you can be on the air talking through repeaters, transmitting via satellites to work other stations thousands of miles away, and even getting a taste of some ionospheric skywave worldwide contacts.

And once you get started, you won't want to stop at Technician! Everyone will want the General Class license that gives you worldwide privileges on every single ham band. To study for your Element 3 exam, use my *General Class* book. On February 23, 2007, a new Federal Communications Commission ruling eliminated the Morse code test requirement for all classes of amateur radio licenses. The same ruling also granted Technician-Plus frequency privileges to no-code Technicians, giving you access to portions of the 75/80-, 40-, 15-, and 10-meter worldwide high frequency bands. So if the dots and dashes have kept you out of ham radio before, there is absolutely no code test required for Technician, General, or Extra class licenses. The only thing you will need to study for your General and Extra class written examinations are my *General Class* and *Extra Class* books!

Three ham radio license classes. No more Morse code test for Technician, General and Extra. It couldn't be easier to get involved in one of the most fascinating and fun hobbies in the world!

Ready to get started? Hurry up – I am regularly on the airwaves, and I hope to make contact with you very soon with your new Technician Class call sign!

Listen to the sounds of ham radio excitement! Play my audio CD included with this book now!!

73!

Gordon West, WB6NOA

About This Book

This book provides you with all of the study materials you need to prepare yourself to take and pass the Element 2 written examination to obtain your Technician Class amateur radio license. Technician Class is the entry-level amateur operator/primary station license issued by the Federal Communications Commission — the FCC. *Absolutely no Morse code test is required* for the Technician Class license, which will give you unlimited VHF and UHF ham band privileges, plus the excitement of operating on worldwide skywave bands, too!

My book also provides you with valuable information you need to be an active participant in the amateur ranks. To help you get the most out of *Technician Class,* here's a look at how my book is organized:

- *Chapter 1* provides an overview of the amateur service and a quick look at all of the exciting things you can do with just your entry-level Technician Class license.
- *Chapter 2* tells you about all of the ham radio operating privileges you will have with your new Technician Class license — including the additional high frequency privileges you will earn when you pass your Technician Class exam.
- *Chapter 3* gives an overview of the amateur service, and a brief history of ham radio regulations. It contains details on the December, 1999 FCC Report & Order that greatly simplified the Amateur Radio service licensing structure, which streamlined the number of examination elements and reduced the emphasis on Morse code for all classes of ham radio licenses. It's your orientation to ham radio.
- *Chapter 4* describes the Element 2, Technician Class written examination, and contains all 394 middle-school-reading-level questions that comprise the Element 2 question pool. Thirty-five of these multiple choice questions will be on your written examination. If you answer 26 of them correctly (74 percent), you will pass the exam and receive your FCC license. My book and companion audio course relate every question in the pool to the real world of operating and getting on the air with ham radio!
- *Chapter 5* will tell you what to expect when you take your Element 2 written exam, where to find an exam session, how you will apply for and receive your new FCC license, and more — all the details you need to know to get your license and get on the air.
- *Chapter 6* talks about Morse code. Even though you don't need the Morse code to become a ham operator, learning code with my audio course is fun, and all new ham operators should learn the code as a basic worldwide language of dots and dashes that will live on forever – especially in emergencies.
- The *Appendix* has valuable lists and reference information. And reading the *Glossary* will get you up to speed on some of the amateur radio "lingo" that you might not understand as you begin studying my book.

1

Getting Into Ham Radio

WELCOME TO AMATEUR RADIO!

There are many great stories about the origin of our "ham" nickname, and soon you will decide on your favorite. Many people call us "hams" because we always seem to be ready to "show off" the magic of our little wireless gadgets. It won't be long before you'll be doing this, too!

Amateur radio operators INVENTED wireless communications. Did you know that Guglielmo Marconi – who often is credited with inventing radio – considered himself an amateur? Amateur operators were 100 years ahead of your little cell phone. Our century-old Morse code dots and dashes were a forerunner to that little PDA that's tucked away in your pocket or purse. Just ask any ham operator who knows his history, and you'll be told that this new hobby and service you are preparing to join truly did shape everything that is going on with wireless communications today and in the future.

There's a lot of fun to be had in the ham radio hobby you're about to join!

Nearly every country in the world has an amateur radio service, and we all share certain ham radio frequency bands globally. There are more than 2,000,000 licensed amateur radio operators throughout the world, and we number more than 650,000 here in the U.S.

WHY DO I NEED A LICENSE?

Unlike a lot of electronic communications devices that you already use – like cordless phones in your home, cell phones on the road, or short-range FRS or CB radios – the ham radio equipment you will learn to use has capabilities to communicate across town, around the world, and even into outer space. So, in order to keep things orderly, all hams are required to demonstrate that they understand the rules, regulations, and international frequency assignments placed on amateur radio.

All countries require their ham radio operators to be licensed; to know their country's local radio rules and regulations, and to know a little about how radios work in order to pass that "entrance exam." Here in the U.S., the exam for the entry-level Technician Class amateur operator license is a snap. Chapter 4 gives you all the details!

Once you pass our 35-question multiple choice "entrance exam," you'll be issued your first amateur radio license by the FCC (Federal Communications Commission). Absolutely no knowledge of Morse code is required for your Technician Class radio license, or any FCC ham license.

Your brand new Technician Class license will authorize you to operate with unrestricted access on all ham bands above 50 MHz, including exciting voice privileges on the long range 10 meter band, plus 3 other worldwide bands for Morse code! Ham "bands" are internationally-designated groups of frequencies reserved for amateur radio operation. In the next Chapter, and throughout this book, you'll learn about frequencies in cycles per second (hertz), ham bands in wavelengths, and exactly where our ham bands are located on the radio dial.

YOUR FIRST RADIO

The first radio that most Technician Class operators start out with is a dual-band handheld transmitter/receiver. You'll tune into frequencies that are automatically relayed to other ham radio operators throughout your city via repeaters. The keypad on the front of your little handheld may also be used to dial up internet radio links. Imagine walking down the street with your handheld talking to a fellow ham radio operator in Australia – or maybe in Antarctica! Or how about calling home on a free ham radio "autopatch" network – or listening to freeway traffic reports that are more frequent and accurate than those on your car radio!

You'll probably want to get a dual-band handheld transceiver for your first ham radio.

Many dual-band handheld ham radios also have the capability to tune in worldwide shortwave broadcasts, AM and FM radio stations, television audio, scan the police, paramedics, and fire frequencies, and even do a little eavesdropping backstage on wireless microphone frequencies. These are very sophisticated pieces of radio equipment with many fun and interesting capabilities designed in.

Later, after you've learned how to operate properly and are ready to expand into other ham radio capabilities, you can consider buying a mobile radio that can be mounted in your car or pickup, and even used as a "base station" in your home. But my strong recommendation for your first radio is the dual-band handheld.

After you've developed your operating skills, you may want a mobile set for your car or pickup. They can be used as base stations in your house, too.

What about a radio tower? Well, while the Eiffel Tower in Paris was one of the world's first antenna towers, today's antennas for ham radio can be very discreet. They can be safely placed in attics, or be stealth hookups to a nearby window screen. So, no, you won't need a huge tower to get started in your new hobby and service.

WHAT ELSE CAN YOU DO? PLENTY!

Your new handheld can plug into a global positioning system (GPS) and relay your location to ham satellites or mountaintop digital repeaters tied into the internet. Your fellow hams – even mom – could track your every movement. Or maybe you'd like to plug your handheld into a tiny color video sender and show all the hams around you what it looks like from the top of that mountain you just climbed!

Your new Technician Class license allows you every ham radio privilege on all of the VHF and UHF bands. Imagine talking thousands of miles away by bouncing 6-meter band signals off of the ionosphere. Six-meter skywave excitement occurs summer and fall, and many Technician operators have worked hundreds of other ham stations throughout the world on 6-meter single sideband. You also gain long-range skywave privileges on a portion of 10 meters, 15 meters, 40 meters, and 80 meters, too!

Are you into radio control (R/C) of model airplanes and boats? There are channels reserved only for use by licensed ham radio operators for this type of activity on 6-meters, too!

What else can you do with all of your new Technician Class privileges? You could set up your 2-meter ham station to bounce signals off the face of the Moon! Or send a digital stream of information off a meteor trail. Or yak with a pal on the other side of the country through one of the many amateur radio satellites orbiting up there for exclusive use by hams. And for out-of-the-world communications with your new 2-meter ham band Technician Class privileges, you can speak regularly with the ham radio operator astronauts passing 240 miles overhead in the International Space Station. Yes, the ISS has a complete ham station installed and, when it's overhead, its within line-of-sight range of your handheld!

There's still more you can do with just that entry-level Technician Class license.

If you're also into computers, hams have access to hundreds of wireless frequencies to send high speed information "packets" over the airwaves, absolutely-error-free wireless e-mails, digital slow scan color photos, and a relatively new binary phase-shift-keying computer-to-computer mode that occupies only a sliver of radio bandwidth called PSK-31.

Live television? As a Technician Class ham radio operator, we can put you on some frequencies where you will join fellow amateur radio television operators to beam crystal clear, live-action television all over the state! We call this "ATV."

Imagine talking on your little handheld radio to another ham with their handheld radio *halfway around the world!* Very common these days, thanks to seasoned ham operators who offer free Voice-Over-Internet-Protocol gateway station access. Both IRLP and EchoLink® gateway stations are standing by to relay your radio call.

If you're non-technical, we can use a computer to download all the frequencies to your handheld radio memory circuits to turn you into a walkin' talkin' computin' radio operator with equipment not much larger than your favorite cell phone. You don't have to be an engineer to become a ham operator!

Not just a hobby, but a service

There's a serious side to our hobby, when ham radio becomes a public service. When emergencies strike, ham operators are at their shining best. At any major or local disaster, ham operators are often the first to handle emergency calls.

Our ham radio nets stay on the air through hurricanes, during tornadoes, and even in the event of major disasters. Following the 9/11 attacks and the Hurricane Katrina disaster, ham operators worked for more than a month providing additional emergency communications capabilities to the rescue workers. More recently, a team of US hams volunteered for more than 2 months to provide emergency communications support following the devistating earthquake in Haiti. Our ham radio network of mobile radios, relay stations and remote base equipment continuously keeps emergency responders in contact with many necessary resources.

Many of our ham radio emergency traffic handlers are always at home and always on the air. You wouldn't know that they are visually impaired or perhaps confined to a wheelchair because there is no disability that would keep ham radio operators from working on the amateur radio service airwaves. To learn more, visit **www.handiham.org.**

Join a club – get yourself an "Elmer"

When you pass your upcoming Technician Class ham exam, your local ham radio clubs may send you a letter with a warm welcome inviting you to join them at an upcoming club meeting. You should go!

Club members – who are now your fellow amateur operators – can help you select radio equipment, program your new radio, and even come to your house and to help set-up that home or vehicle radio installation. These willing helpers who are eager to "show you the ropes" are known as "Elmers," and your Elmer can teach you the practical, on-the-air aspects of amateur radio. Ham radio is one big fraternity!

What are some other ways to learn about your new hobby? One of the best is to read some of the ham radio magazines. One that focuses exclusively on Technician Class frequencies if ***CQ VHF–Ham Radio Above 50 MHz*** (visit **www.cq-vhf.com**). Their parent magazine, ***CQ – Amateur Radio*** contains excellent articles that will help you learn, as well. There's a coupon in the back of this book for a free, 3-issue mini-subscription to ***CQ***.

The largest ham radio magazine is ***QST – Amateur Radio***. It is published by the American Radio Relay League (ARRL), which is the national association for amateur radio in the United States (visit **www.arrl.org**). There's a membership application form to join the ARRL in the back of this book, too, which will also entitle you to a free *Repeater Directory* or other ARRL publication.

And before I let you go on to the next Chapter of my book – which details all of the Technician Class frequency privileges you'll earn when you pass your exam – keep in mind that there are more frequencies and bands that you will earn as you upgrade your ham license. As a General and Extra class operator, you'll gain many more long-range bands to keep you in touch around the world. You could even become a volunteer examiner with the General and Extra class licenses, and then you could give the same exam you are about to pass.

So let's get started now with Technician Class study. In the next Chapter, we'll take an in-depth look at the frequency privileges you'll earn with your Technician Class license. I can't wait to hear you on the air with your new call sign. Welcome to our Amateur Radio service!

2

Technician Class Privileges

There is plenty of excitement out there on the amateur VHF and UHF bands for the ham with a Technician Class license. And thanks to the FCC's ruling that eliminated the Morse code test requirement for all classes of ham radio license, the Technician Class operator also gets a taste of HF excitement on portions of the 75/80-, 40-, 15-, and 10-meter bands. Some of these band segments are restricted to Morse code only operation, so you'll still want to learn the code, even though it isn't required by the Federal Communications Commission.

SPECTRUM, WAVELENGTH, AND FREQUENCY

Before we look at the actual Technician Class privileges you'll earn with your new FCC Amateur Radio license, let's take a minute to understand the fundamentals of what is meant by the radio terms *spectrum, wavelength,* and *frequency.*

Figure 2-1 on the next page shows the entire electromagnetic energy spectrum, and highlights where the radio frequencies fall within the total spectrum. The low end of the spectrum starts with audio and VLF (Very Low Frequency) frequencies. At the top end of the overall spectrum – above the radio frequencies – are light, X-Rays, and Gamma Rays.

The region from 20,000 hertz to 30 gigahertz is where radio waves are found. Within that region, the radio spectrum is divided up for various uses. The commercial radio AM band is found from 550 kHz (kilohertz) to 1650 kHz. FM radio stations operate between 88 MHz (megahertz) to 108 MHz. One hertz is equal to one cycle per second. That means that an AM signal at 720 kHz on you radio dial is oscillating at 720,000 cycles per second, and an FM signal at 91.5 MHz on your dial is oscillating at 91,500,000 cycles per second!

Figure 2-2 adds some detail to this explanation. The table at the top shows where some of the frequencies lie, and where Amateur Radio operators have privileges. The illustrations show how wavelength and frequency are related. The easiest thing to remember is LOWER LONGER, HIGHER SHORTER. Lower frequency radio waves travel longer distances in one cycle (wavelength), and higher frequency radio waves travel shorter distances in one cycle.

When we say that we are going to operate on the 6-meter band, that means the wavelength of the frequency we will be using is about 6-meters long – or that one cycle of the frequency travels about 236 inches (or 19 feet) in one cycle. On the 70 centimeter band, one wavelength is about 27.5 inches long (or about 2.3 feet) in one cycle.

In general, antennas need to be equal to wavelength (or ½ or ¼ wavelength) in order to efficiently send and receive radio signals. And, finally, radio waves travel at approximately the speed of light, or 300,000,000 meters per second. So, there's a lot of stuff happening in a big hurry out there on the radio waves!

Now, let's take a look at the Technician Class frequency privileges, and how they are used for various purposes.

Figure 2-1. The electromagnetic spectrum detailing the radio frequency spectrum
Soruce: FCC

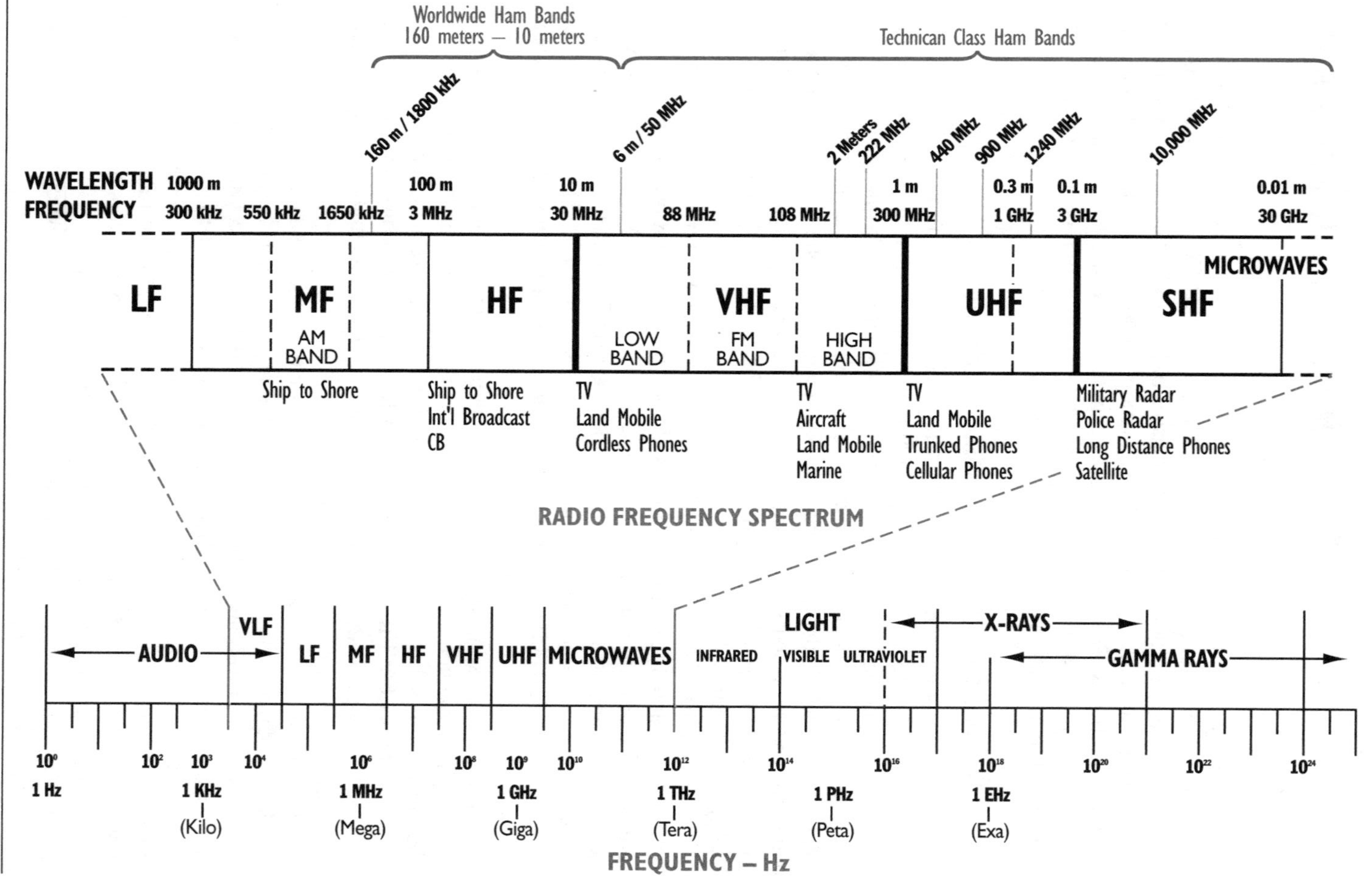

Figure 2-2. Radio Bands, Frequency and Wavelength

Category	Abbrev.	Frequency	Amateur Band Wavelength
Audio	AF	20 Hz to 20 kHz	None
Very Low Frequency	VLF	3 to 30 kHz	None
Low Frequency	LF	30 to 300 kHz	None
Medium Frequency	MF	300 to 3000 kHz	160 meters
High Frequency	HF	3 to 30 MHz	80, 40, 30, 20, 17, 15, 12, 10 meters
Very High Frequency	VHF	30 to 300 MHz	6, 2, 1.25 meters
Ultrahigh Frequency	UHF	300 to 3000 MHz	70, 33, 23, 13 centimeters
Superhigh Frequency	SHF	3 to 30 GHz	9, 5, 3, 1.2 centimeters
Extremely High Frequency	EHF	Above 30 GHz	6, 4, 2.5, 2, 1 millimeter

Frequency Spectrum

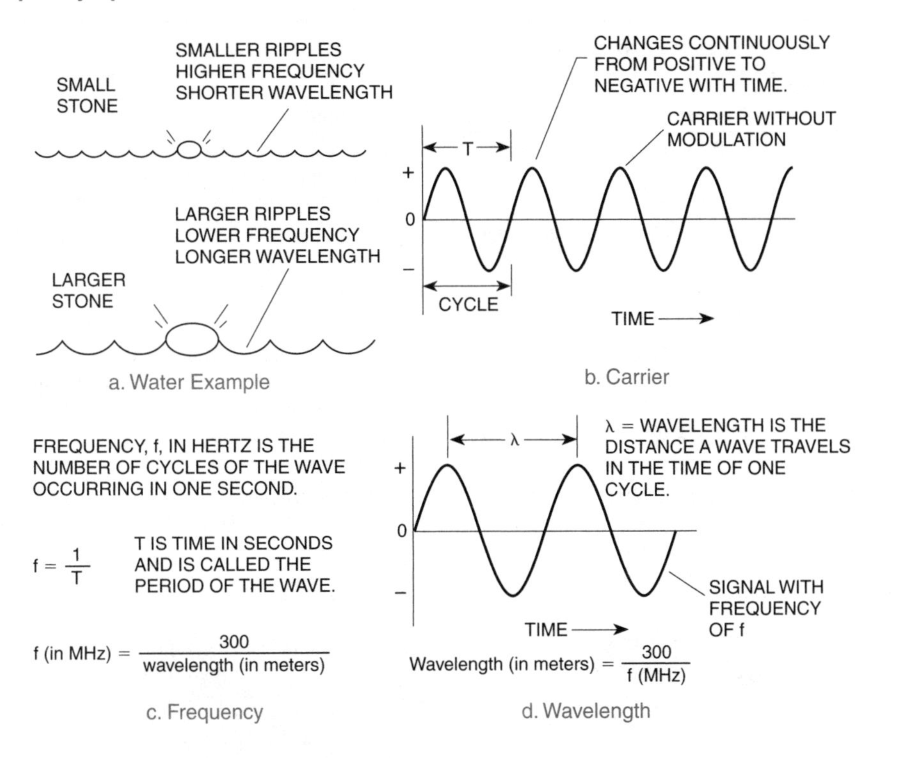

Carrier, Frequency, Cycle and Wavelength

TECHNICIAN CLASS PRIVILEGES

The Technician Class is the entry-level license where you get started in the amateur service. It's now easier than ever to enter the amateur radio service as a Technician Class operator.

As a Technician Class operator, you will have plenty of excitement on radio frequencies above 30 MHz. You will have full operating privileges on all of the exciting VHF, UHF, SHF, and microwave bands shown in *Table 2-1.* You can work skywaves on 6 meters to communicate all over the country; on the 2-meter band, you'll operate through repeaters and orbiting satellites, and on voice over Internet links. On the 222 MHz band, you may operate through linked repeaters; on the 440 MHz band, you might try amateur television, satellite, and remote base operation; and on 1270 MHz there are more frequencies for amateur television, satellites, repeater linking, and digital messaging and voice over Internet linking. And then there are the microwave bands where dish, loop, and Yagi antennas will beam out signals a lot farther than you might think.

In adddition, you also may operate single sideband voice on a portion of the 10 meter band, where skywaves bounce halfway around the world. You also receive Morse code privileges on portions of the 10-, 15-, 40-, and 80-meter bands. While it's not a requirement to learn the code, or pass a code test, you may enjoy Morse code worldwide operation with your new Technician Class privileges on these 4 worldwide bands.

Table 2-1. Technician Class VHF/UHF Unrestricted Operating Privileges

Wavelength Band	Frequency	Emissions	Comments
6 Meters	50.0–54.0 MHz	All modes	Sideband voice, radio control, FM repeater, digital computer, remote bases, and autopatches. Even CW. (1500 watts PEP output)
2 Meters	144–148 MHz	All modes	All types of operation including satellite and owning repeater and remote bases. (1500 watt PEP output)
1¼ Meters	222–225 MHz	All modes	All band privileges. (1500 watt PEP output)
70 cm	420–450 MHz	All modes	All band privileges, including amateur television, packet, Internet linking, FAX, and FM voice repeaters. (1500 watt PEP output.)
33 cm	902–928 MHz	All modes	All band privileges. Plenty of room! (1500 watt PEP output.)
23 cm	1240–1300 MHz	All modes	All band privileges. (1500 watt PEP output)
13 cm	2300–2310 MHz 2390–2450 MHz	All modes	Amateur television

Kids make great ham radio operators.

6-METER WAVELENGTH BAND, 50.0-54.0 MHz

The Technician Class operator will enjoy all amateur service privileges and maximum output power of 1500 watts on this worldwide band. Are you into radio control (R/C) and want to escape the interference between 72 and 76 MHz? On 6 meters, your Technician Class license allows you to operate on exclusive radio control channels at 50 MHz and 53 MHz, just for licensed hams. *Table 2-2* shows the ARRL 6-meter wavelength band plan.

On 6 meters, the Technician Class operator can get a real taste of long-range skywave skip communications. During the summer months, and during selected days and weeks out of the year, 50-54 MHz, 6-meter signals are refracted by the ionosphere, giving you incredible long-range communication excitement. It's almost a daily phenomenon during the summer months for 6 meters to skip all over the country. This is the big band for the Technician Class operator because of this type of ionospheric, long-range, skip excitement. There are even repeaters on 6 meters. So make 6 meters "a must" at your future operating station.

Table 2-2. 6-Meter Wavelength Band Plan, 50.0-54.0 MHz

MHz	Use
50.000–50.100	CW weak signal
50.060–50.080	CW Beacon FM
50.100–50.300	SSB, CW
50.100–50.200	DX window & SSB DX calling
50.225	Domestic SSB calling frequency and QSO each side
50.300–50.600	Non-voice communications
50.620	Digital/Packet calling frequency
50.800–50.980	Radio control
	20 kHz channels
51.000–51.100	Pacific DX window
51.120–51.480	Repeater inputs (19)
51.120–51.180	Digital repeater inputs
51.620–51.980	Repeater outputs (19)
51.620–51.680	Digital repeater outputs
52.000–52.480	Repeater inputs (23)
52.020, 52.040	FM simplex
52.500–52.980	Repeater outputs (23)
52.525, 52.540	FM simplex
53.000–54.480	Repeater inputs (19)
53.000, 53.020	FM simplex
53.1/53.2/53.3/53.4	Radio control
53.500–53.980	Repeater outputs (19)
53.5/53.6/53.7/53.8	Radio control
53.520	Simplex
53.900	Simplex

2-METER WAVELENGTH BAND, 144-148 MHz

The 2-meter band is the world's most popular spot for staying in touch through repeaters. Here is where most all of those hand-held transceivers operate, and the Technician Class operator receives unlimited 2-meter privileges! *Table 2-3* gives the 2-meter wavelength band plan adopted by the ARRL VHF/UHF advisory committee.

The United States, and many parts of the world, are blanketed with clear, 2-meter repeater coverage. They say there is nowhere in the United States you can't reach at least one or two repeaters with a little hand-held transceiver. 2-meters has you covered! Here are examples:

Handie-talkie channels	Tropo-DX-ducting	Remote base
Transmitter hunts	Internet radio links	Simplex operation
Autopatch	Contests	Rag-chewing
Moon bounce (EME)	Traffic handling	Emergency nets
Meteor bursts	Satellite downlink	Sporadic-E DX
Packet radio	Satellite uplink	Aurora

The Technician Class license allows 1500 watts maximum power output for specialized 2-meter communications, and also permits you to own and control a 2-meter repeater.

Table 2-3. ARRL 2-Meter Wavelength Band Plan, 144-148 MHz

MHz	Use
144.00–144.05	EME (Earth-Moon-Earth) (CW)
144.05–144.06	Propagation beacons (old band plan)
144.06–144.10	General CW and weak signals
144.10–144.20	EME and weak-signal SSB
144.20	National SSB calling frequency
144.20–144.275	General SSB operation, upper sideband
144.275–144.30	New beacon band
144.30–144.50	New OSCAR subband plus simplex
144.50–144.60	Linear translator inputs
144.60–144.90	FM repeater inputs
144.90–145.10	Weak signal and FM simplex plusD-Star digital
145.10–145.20	Linear translator outputs plus packet
145.20–145.50	FM repeater outputs
145.50–145.80	Miscellaneous and experimental modes
145.80–146.00	OSCAR subband—satellite use only
146.01–146.37	Repeater inputs
146.40–146.58	Simplex
146.61–146.97	Repeater outputs
147.00–147.39	Repeater outputs
147.42–147.57	Simplex and D-Star systems
147.60–147.99	Repeater inputs

1¼-METER WAVELENGTH BAND, 219-220 MHz & 222-225 MHz

On the 222 MHz band, the frequencies 219 MHz to 220 MHz may be used by point-to-point digital message forwarding stations operated by Technician Class licensees or higher. These stations must register with the American Radio Relay League 30 days prior to activation, and must not interfere with primary marine users near the Mississippi River, or any other primary user of this band. Remember, before you turn on a point-to-point digital message forwarding station, you must first register your operation with the ARRL 30 days before going on the air on 219 to 220 MHz. The 100 kHz channels for point-to-point fixed digital message forwarding stations, 50 watts PEP limit, are shown in *Table 2-4.*

Table 2-4. ARRL 1¼-Meter Wavelength Band Plan, 219-220 MHz

MHz	Use			
219–220	Point-to-point fixed digital message forwarding systems. Must be coordinated through ARRL. 100 kHz Channels. 50W PEP limit.			
	Channel	**Freq. (MHz)**	**Channel**	**Freq. (MHz)**
	A	219.050	F	219.550
	B	219.150	G	219.650
	C	219.250	H	219.750
	D	219.350	I	219.850
	E	219.450	J	219.950
220–222	No longer available			

Table 2-5 shows the usage allocation for this portion of the band. The Technician Class license permits you to use the entire band at 1500 watts maximum output power. If you need some relief from the activity on 2 meters, the 222-225 MHz band is similar in propagation and use. 222 MHz to 225 MHz is now exclusively assigned to our Amateur Radio service.

Table 2-5. ARRL 1¼-Meter Wavelength Band Plan, 222-225 MHz

MHz	Use
222.00–222.15	Weak-signal modes (FM only)
222.00–222.05	EME (Earth-Moon-Earth)
222.05–222.06	Propagation beacons
222.10	SSB and CW calling frequency
222.10–222.15	Weak signal CW and SSB
222.15–222.25	Local coordinator's option: Weak signal, ACSB, repeater inputs, control points
222.25–223.38	FM repeater inputs only
223.40–223.52	FM simplex
223.50	Simplex calling frequency
223.52–223.64	Digital, packet
223.64–223.70	Links, control
223.71–223.85	Local coordinator's option: FM simplex, packet, repeater outputs
223.85–224.98	Repeater outputs only

70-CM WAVELENGTH BAND, 420-450 MHz

As you gain more experience on the VHF and UHF bands, you will soon be invited to the upper echelon of specialty clubs and organizations. The 450-MHz band is where the experts hang out. *Table 2-6* presents the ARRL 70-cm (centimeter) wavelength band plan. Amateur television (ATV) is very popular, so there's no telling who you may see as well as hear. This band also has the frequencies for controlling repeater stations and base stations on other bands, plus satellite activity. With a Technician Class license, you may even be able to operate on General Class worldwide frequencies if a General Class or higher control operator is on duty at the base control point. You would be able to talk on your 450-MHz hand-held transceiver and end up in the DX portion of the 20-meter band. As long as the control operator is on duty at the control point, your operation on General Class frequencies is completely legal!

The 450 MHz band is also full of packet communications, RTTY, FAX, and all those fascinating FM voice repeaters. If you are heavy into electronics, you'll hear fascinating topics discussed and digitized on the 450-MHz band. A Technician Class operator has full power privileges as well as unrestricted emission privileges. Visit: **www.wa6twf.com** and **www.winsystem.org**

Table 2-6. ARRL 70-cm Wavelength Band Plan, 420-450 MHz

MHz	Use
420.00–426.00	ATV repeater or simplex with 421.25-MHz video carrier control links and experimental
426.00–432.00	ATV simplex with 427.250-MHz video carrier frequency
432.00–432.07	EME (Earth-Moon-Earth)
432.07–432.08	Propagation beacons (old band plan)
432.08–432.10	Weak-signal CW
432.10	70-cm calling frequency
432.10–433.00	Mixed-mode and weak-signal work
432.30–432.40	New beacon band
433.00–435.00	Auxiliary/repeater links
435.00–438.00	Satellite only (internationally)
438.00–444.00	ATV repeater input with 439.250-MHz video carrier frequency and repeater links
442.00–445.00	Repeater inputs and outputs (local option)
445.00–447.00	Shared by auxiliary and control links, repeaters and simplex (local option); (446.0-MHz national simplex frequency)
447.00–450.00	Repeater inputs and outputs

33-CM WAVELENGTH BAND, 902-928 MHz

Radio equipment manufacturers are just beginning to market equipment for this band. Many hams are already on the air using home-brew equipment for a variety of activities. If you are looking for a band with the ultimate in elbow room, this is it! *Table 2-7* shows the 33-cm wavelength band plan adopted by the ARRL.

Table 2-7. ARRL 33-cm Wavelength Band Plan, 902-928 MHz

MHz	Use
902.0–903.0	Weak signal (902.1 calling frequency)
903.0–906.0	Digital Communications (903.1 alternate calling frequency)
906.0–909.0	FM repeater inputs
909.0–915.0	ATV
915.0–918.0	Digital Communications
918.0–921.0	FM repeater outputs
921.0–927.0	ATV
927.0–928.0	FM simplex and links

23-CM WAVELENGTH BAND, 1240-1300 MHz

There is plenty of equipment for this band. Technician Class operators may run any legal amount of power—with 20 watts about the usual safe limit. The frequencies are in the microwave region, and this band is excellent to use with local repeaters in major cities. Like the 450-MHz and the 2-meter bands, this band is sliced into many specialized operating areas. You can work orbiting satellites, operate amateur television, or own your own repeater with your Technician Class license. *Table 2-8* presents the 23-cm wavelength band plan adopted by the ARRL.

Table 2-8. ARRL 23-cm Wavelength Band Plan, 1240-1300 MHz

MHz	Use
1240–1246	ATV #1
1246–1248	Narrow-bandwidth FM point-to-point links and digital, duplexed with 1258-1260 MHz
1248–1252	Digital communications
1252–1258	ATV #2
1258–1260	Narrow-bandwidth FM point-to-point links and digital, duplexed with 1246-1252 MHz
1260–1270	Satellite uplinks, reference WARC '79
1260–1270	Wide-bandwidth experimental, simplex ATV
1270–1276	Repeater inputs, FM and linear, paired with 1282-1288 MHz, 239 pairs every 25 kHz, e.g., 1270.025, 1270.050, 1270.075, etc. 1271.0-1283.0 MHz uncoordinated test pair
1276–1282	ATV #3
1282–1288	Repeater outputs, paired with 1270-1276 MHz
1288–1294	Wide-bandwidth experimental, simplex ATV
1294–1295	Narrow-bandwidth FM simplex services, 25-kHz channels
1294.5	National FM simplex calling frequency
1295–1297	Narrow bandwidth weak-signal communications (no FM)
1295.0–1295.8	SSTV, FAX, ACSB, experimental
1295.8–1296.0	Reserved for EME, CW expansion
1296.0–1296.05	EME exclusive
1296.07–1296.08	CW beacons
1296.1	CW, SSB calling frequency
1296.4–1296.6	Crossband linear translator input
1296.6–1296.8	Crossband linear translator output
1296.8–1297.0	Experimental beacons (exclusive)
1297–1300	Digital communications

10-GHz (10,000 MHz!) BANDS AND MORE

There are several manufacturers of ham microwave transceivers and converters for this range, so activity is excellent. Gunnplexers are the popular transmitter. Using horn and dish antennas, 10 GHz is frequently used by hams to establish voice communications for controlling repeaters over paths from 20 miles to 100 miles. Output power levels are usually less than one-eighth of a watt! It's really fascinating to see how directional the microwave signals are. If you live on a mountain-top, 10 GHz is for you.

All modes and licensees except Novices are authorized on the bands shown in *Table 2-9.* There is much Amateur Radio experimentation on these bands.

Table 2-9. Gigahertz Bands

2.30–2.31 GHz	10.0–10.50 GHz*	119.98–120.02 GHz
2.39–2.45 GHz	24.0–24.25 GHz	142.0–149.0 GHz
3.30–3.50 GHz	47.0–47.20 GHz	241.0–250.0 GHz
5.65–5.925 GHz	75.50–81.0 GHz	All above 300 GHz

*Pulse not permitted

HIGH-FREQUENCY (HF) TECHNICIAN CLASS PRIVILEGES

In the previous pages of this Chapter, we described in detail all of your Technician Class privileges above 30 MHz. Now we want to tell you about the additional HF band privileges you have as a Technician since the FCC eliminated the Morse code test requirement. You'll see that many of these lower frequencies are restricted to CW operation only. While you aren't required to pass a code test, if you want to operate code on these bands, we strongly encourage you to learn Morse code *before* you get on these frequencies as a new ham operator. You'll hear lots of experienced CW operators on these bands, and if you try getting on without knowing at least a little bit of code you'll embarrass yourself and you'll annoy the heck out of the experienced hams. So learn the code!

75/80-METER WAVELENGTH BAND, 3500-4000 kHz

Your privileges on the 75/80-meter band are for CW only from 3525 kHz to 3600 kHz.

40-METER WAVELENGTH BAND, 7000-7300 kHz

You will have Morse-code-only privileges on this band from 7025 kHz to 7125 kHz. This is a popular night-time and early morning band because signals in code can reach up to 5,000 miles away!

15-METER WAVELENGTH BAND, 21,000-21,450 kHz

Technician Class operators may operate Morse code from 21,025 kHz to 21,200 kHz in this portion of the worldwide band. You can expect daytime range in excess of 10,0000 miles using CW.

10-METER WAVELENGTH BAND, 28,000-29,700 kHz

You may operate code and digital computer communications from 28,000 kHz to 28,300 kHz, and monitor 28.2 MHz to 28.3 MHz for low-power propagation beacons. Now here's the good news – you may operate single-sideband voice between 28.3 MHz to 28.5 MHz, and literally work the world during band openings during daylight hours! 10 meters is a great band to introduce you to the excitement of long-range, worldwide voice operation on the HF bands. If you have a CB antenna, it will work very well for Technician Class privileges on 10 meters!

Table 2-10. Novice and Technician Class Skywave Operating Privileges

Wavelength Band	Frequency	Emissions	Comments
80 Meters	3525–3600 kHz	Code only	Limited to Morse code (200 watt PEP output limitation)
40 Meters	*7025–7125 kHz	Code only	Limited to Morse code (200 watt PEP output limitation)
15 Meters	21,025–21,200 kHz	Code only	Limited to Morse code (200 watt PEP output limitation)
10 Meters	28,000–28,300 kHz	Data and code	Morse code (200 watt PEP output code limitation)
	28,300–28,500 kHz	Phone and code	Sideband voice (200 watt code PEP output limitation)
Plus These Existing VHF/UHF Frequency Privileges			
6 Meters	50.0–54.0 MHz	All modes	Morse code, sideband voice, radio control, FM repeater, digital computer, remote bases, and autopatches (1500 watts PEP output)
2 Meters	144–148 MHz	All modes	All types of operation including satellite and owning repeater and remote bases. (1500 watt PEP output)
1¼ Meters	219–220 MHz	Data	Point-to-Point digital message forwarding
	222–225 MHz	All modes	All band privileges. (1500 watt PEP output)
70 cm	420–450 MHz	All modes	All band privileges, including amateur television, packet, RTTY, FAX, and FM voice repeaters. (1500 watt PEP output)
33 cm	902–928 MHz	All modes	All band privileges. Plenty of room! (1500 watt PEP output)
23 cm	1240–1300 MHz	All modes	All band privileges. (1500 watt PEP output)
13 cm	2300-2310 MHz 2390-2450 MHz	All modes	Ham T.V. Links; Satellites

* U.S. licensed operators in other than our hemisphere (ITU Region 2) are authorized 7050-7075 kHz due to shortwave broadcast interference.

If you're joining the ham radio hobby from CB radio operating on 11 meters, you know all the excitement with skywave propagation. Your Technician Class 10 meter privileges allow you to own and operate a big worldwide radio, and enjoy ionospheric skip at its best! As we begin to approach the peak of solar cycle 24, conditions on 10 meters will constantly surprise the daylights out of you!

But I *encourage* you to learn CW, even if it isn't a requirement any longer! Ham radio operators all recognize the importance of knowing Morse code. In an emergency, Morse code is that universal language that requires only a minimum of equipment to send and receive messages. In fact, ham radio rescuers during 9/11 saved several trapped firemen sending Morse code by tapping SOS on a metal pipe. United States Coast Guard helicopter crews were able to read CW relief messages beamed up via flashlights during Katrina rescue efforts! And during regular ham radio operating events, Morse code can make a contact much easier than trying to compete a call just using your voice.

Learning Morse code is much like a ham radio initiation – and a tradition that reaches back to the beginnings of our hobby and service. Chapter 6 of this book has a complete code-learning lesson plan that accompanies the many Morse code audio training materials I have prepared, available at the same dealer who sold you this book.

PLAY THE ENCLOSED CD NOW!

The CD enclosed inside the front cover of this book is your introduction to the excitement that awaits you when you get on the air with your new Technician Class privileges.

Track 1 introduces you to the worldwide high frequency bands that your Technician license authorizes. Talk thousands of miles using skip communications on the 10 meter band.

Track 2 introduces you to skywave contacts you can achieve on the 6 meter band.

Track 3 lets you listen in on everything the 2 meter band has to offer the new Technician Class operator. You can even talk to the International Space Station astronauts on 2 meters!

Track 4 puts you on the 222 MHz band for special event stations and more.

Track 5 tunes you in to some exciting satellite communications, plus ham color TV!

Track 6 takes you to the top bands for Technician Class radio excitement. Imagine communicating on frequencies 10 times higher that your microwave oven!

The CD plays on a standard automobile or boom-box player, with no need to pull out the computer. I recorded this CD to acquaint you with all of the sounds of the airwaves, and everything you hear on the CD is something you will be able to do with your new Technician Class license!

If you like the sound of this kind of Technician Class excitement, you can learn even more about the sounds you'll be hearing over the airwaves with a copy of *VHF Propagation – A Practical Guide for Radio Amateurs* by Ken Neubeck, WB2AMU. The book includes another CD recorded by Ken and me, and it gives you in-depth advice on how to operate long-distance on VHF frequencies and discusses how

Gordo and Ken Neubeck explain 6-meter propagation to a ham class

ionospheric propagation works. Write me at WB6NOA@arrl.net for information on how you can obtain your copy of this book and the CD.

And for those of you looking to operate primarily on the "Magic Band" of 6-Meters, there is a book dedicated solely to this band written by Ken Neubeck, *Six Meters – A Guide to the Magic Band.* This book also is available by writing me at WB6NOA@arrl.net.

LEARNING MORE ABOUT HAM RADIO

How do you get to be a knowledgeable amateur radio operator – one who understands how radio works and the ins-and-outs of our hobby? As I mentioned at the end of Chapter 1, one of the best ways to learn all about your new avocation is by reading the amateur radio magazines. One of my favorites is ***CQ – Amateur Radio***. Every month, ***CQ*** is full of a range of articles that will give you operating tips, show you how radio technology works, looks at new products, and always includes a column, Beginner's Corner, to help you get started in ham radio.

And to help you get started with ***CQ***, we've included a special offer in the back of this book. It is a coupon good for a 3-issue, mini-subscription to ***CQ*** – that's **FREE** 3 issues of the magazine to help you get started in ham radio. All you have to do is clip out the order form, fill in your name and address, and drop it in a nearby mailbox. So give ***CQ*** a try. There's a lot of good info in it every month!

AN IMPORTANT WORD ABOUT SHARED FREQUENCIES

In this Chapter, we discussed the Amateur Radio frequency privileges you will receive when you pass your exam and receive your license and call letters from the FCC. It is important, however, that you know that every ham band above 225 MHz is shared on a secondary basis with other services. This means that the primary users get first claim to the frequency!

For example, Government radiolocation (RADAR) is a primary user of some bands. And a multitude of industrial, scientific and medical services have access to the 902-928 MHz band. Just because the frequency is allocated to the amateur service does not mean that others do not have prior right or an equal right to the spectrum. You must not interfere with other users of the band.

There also are instances where amateurs must not cause interference to other stations, such as foreign stations operating along the Mexican and Canadian borders, military stations near military bases, and FCC monitoring stations. Also, amateur operators must not cause interference in the so-called National Radio Quiet Zones which are near radio astronomy locations. The astronomy locations are protected by law from Amateur Radio interference. Operation aboard ships and aircraft also is restricted. The FCC also can curtail the hours of your operation if you cause general interference to the reception of telephone or radio/TV broadcasting by your neighbors.

Every amateur should have a copy of the Amateur Radio Service Part 97 Rules and Regulations. It would be good for you to especially read Part 97.303 on frequency sharing.

You can obtain a copy of the current Part 97 Rules
from the W5YI Group by calling 1-800-669-9594,
or on-line at www.w5yi.org

SUMMARY

There's a world of excitement waiting for you when you pass the Technician Class exam!

Enjoy the excitement of being a ham radio operator and communicate worldwide with other amateur operators. Your Technician Class operator license allows you all ham operator privileges on all bands with frequencies greater than 50 MHz. You have operating privileges on the worldwide 6-meter band, on the world's popular 2-meter and 222-MHz repeater bands, on the amateur television, satellite communications, and repeater-linking 440-MHz and 1270-MHz bands, and on the line-of-sight microwave bands at 10 GHz and above.

Remember, you may enter the amateur service as a Technician Class operator just by passing the Element 2 written exam. You will receive full privileges on all of the VHF/UHF bands we have just described, along with the HF priviliges on 75/80-, 40-, 15-, and 10-meters.

MOST IMPORTANT, ask your examination team members how to join the local amateur radio club. Fellow ham members will welcome you to our fraternity, and give you a hand programming your new radio equipment.

Get on the air! Don't even think about upgrading to the next level of ham licensing until you have become an active ham radio operator on the airwaves.

All of these questions and answers you are about to study will begin to make sense once you begin operating on the air! A simple dual-band handheld, covering 2 meters and 70 cm, will pull in exciting radio communications that you will certainly enjoy. You could even talk with an astronaut, too, aboard the International Space Station with that little handheld transceiver!

Now, let's take a look at a little ham radio history – read on.

3

A Little Ham History!

Ham radio has changed a lot in the 100+ years since radio's inception. In the past 25 years, we have seen some monumental changes! So, before we get started preparing for the exam, I'm going to give you a little history lesson about our hobby, its history, and an overview of how you'll progress through the amateur ranks from your first, entry-level Technician Class license to the top amateur ticket – the Amateur Extra Class license. We know this background knowledge will make you a better ham! I'll keep it light and fun, so breeze through these pages.

In this chapter you'll learn all of the licensing requirements under the FCC rules that became effective April 15, 2000. And you'll learn about the six classes of license that were in effect *prior* to those rules changes. That way, when you run into a Novice, Technician Plus or Advanced class operator on the air, you'll have some understanding of their skill level, experience, and frequency privileges.

WHAT IS THE AMATEUR SERVICE?

There are more than 650,000 licensed amateur radio operators in the U.S. today. The Federal Communications Commission, the Federal agency responsible for licensing amateur operators, defines our radio service this way:

"The amateur service is for qualified persons of all ages who are interested in radio technique solely with a personal aim and without pecuniary interest."

Ham radio is first and foremost a fun hobby! In addition, it is a service. And note the word "qualified" in the FCC's definition – that's the reason why you're studying for an exam; so you can pass the exam, prove you are qualified, and get on the air.

Millions of operators around the world exchange ham radio greetings and messages by voice, teleprinting, telegraphy, facsimile, and television worldwide. Japan, alone, has more than a million hams! It is very commonplace for U.S. amateurs to communicate with Russian amateurs, while China is just getting started with its amateur service. Being a ham operator is a great way to promote international good will.

The benefits of ham radio are countless! Ham operators are probably known best for their contributions during times of disaster. In recent years, many recreational sailors in the Caribbean who have been attacked by modern-day pirates have had their lives saved by hams directing rescue efforts. Following the 9/11 terrorist attacks on the World Trade Center and the Pentagon, literally thousands of local hams assisted with emergency communications. In addition, over the years, amateurs have contributed much to electronic technology. They have even designed and built their own orbiting communications satellites.

The ham community knows no geographic, political or social barrier. If you study hard and make the effort, you are going to be part of our fraternity. Follow the suggestions in my book and your chances of passing the written exam are excellent.

A BRIEF HISTORY OF AMATEUR RADIO LICENSING

Government licensing of radio stations and amateur operators began with The Radio Act of 1912, which mandated the first Federal licensing of all radio stations and assigned amateurs to the short wavelengths of less than 200 meters. These "new" requirements didn't deter them, and within a few years there were thousands of licensed ham operators in the United States.

Since electromagnetic signals do not respect national boundaries, radio is international in scope. National governments enact and enforce radio laws within a framework of international agreements which are overseen by the International Telecommunications Union. The ITU is a worldwide United Nations agency headquartered in Geneva, Switzerland. The ITU divides the radio spectrum into a number of frequency bands, with each band reserved for a particular use. Amateur radio is fortunate to have many bands allocated to it all across the radio spectrum.

In the U.S., the Federal Communications Commission is the government agency responsible for the regulation of wire and radio communications. The FCC further allocates frequency bands to the various services in accordance with the ITU plan – including the Amateur Service – and regulates stations and operators.

In the early years of amateur radio licensing in the U.S., the classes of licenses were designated by the letters "A," "B," and "C." The highest license class with the most privileges was "A." In 1951, the FCC dropped the letter designations and gave the license classes names. They also added a new Novice class. In 1967, the Advanced class was added to the Novice, Technician, General and Extra Classes. The General exam required 13-wpm code speed, and Extra required 20-wpm. Each of the five written exams were progressively more comprehensive and formed what came to be known as the *Incentive Licensing System.*

In 1979, the international Amateur Service regulations were changed to permit all countries to waive the manual Morse code proficiency requirement for "...stations making use exclusively of frequencies above 30 MHz." This set the stage for the creation of the Technician "no-code" license, which occurred in 1991, when the 5-wpm Morse code requirement for the Technician Class was the eliminated.

By this time, there was a total of six Amateur Service license classes – Novice, Technician, Technician-Plus, General, Advanced, and Extra – along with five written exams and three Morse code tests used to qualify hams for their various licenses.

The Amateur Service Is Restructured

Following an extensive review begun in 1998, the FCC implemented a complete restructuring of the U.S. amateur service that became effective April 15, 2000. Today, applicants can only be examined for three amateur license classes:

- Technician Class – the VHF/UHF entry level license;
- General Class – the HF entry level license, and
- Amateur Extra Class – a technically-oriented senior license.

Most recently, the FCC updated its amateur radio rules by eliminating the Morse code test requirement. The new rule went into effect in February, 2007, and it means that you can earn any class of license – Technician, General, and Extra Class – simply by passing written exams

Individuals with licenses issued before April 15, 2000, have been "grandfathered" under the new rules. This means that Novice and Advanced class amateurs are able to modify and renew their licenses indefinitely. Technician-Plus amateur licenses will be renewed as Technician Class, keeping their HF operating privileges indefinitely. The FCC elected not to change the operating privileges of any class, so you may hear some of these "grandfathered" hams when you get on the air.

Self-Testing In The Amateur Service

Prior to 1984, all amateur radio exams were administered by FCC personnel at FCC Field Offices around the country. In 1984, the VEC (Volunteer Examiner Coordinator) System was formed after Congress passed laws that allowed the FCC to accept the services of Volunteer Examiners (or VEs) to prepare and administer amateur service license examinations. The testing activity of VEs is managed by Volunteer Examiner Coordinators (or VECs). A VEC acts as the administrative liaison between the VEs who administer the various ham examinations and the FCC, which grants the license.

A team of three VEs, who must be approved by a VEC, is required to conduct amateur radio examinations.

In 1986, the FCC turned over responsibility for maintenance of the exam questions to the National Conference of VECs, which appointed a Question Pool Committee (QPC) to develop and revise the various question pools according to a schedule.

That completes your history lesson. If you'd like to learn more, visit The Ham Radio History forum at **www.YahooGroups.com/list/Ham-Radio-History**. Now let's turn our attention to the privileges you'll earn as a new Technician Class operator.

LICENSE PRIVILEGES

An amateur operator license conveys many privileges. As the control operator of an amateur radio station, you will be responsible for the quality of the station's transmissions. Most radio equipment must be authorized by the FCC before it can be widely used by the public but, for the most part, this is not true for amateur equipment!

Unlike the citizen's band service, amateurs may design, construct, modify and repair their own equipment. But you must have a license to do this, and even though it is easier than ever, there are certain things you must know before you can obtain your license from the FCC. Everything you need to know is covered in this book.

OPERATOR LICENSE REQUIREMENTS

To qualify for an amateur operator/primary station license, a person must pass an examination according to FCC guidelines. The degree of skill and knowledge that the candidate demonstrates to the examiners determines the class of operator license for which the person is qualified.

Anyone is eligible to become a U.S. licensed amateur operator (including foreign nationals, if they are not a representative of a foreign government). There is no age limitation – if you can pass the examinations, you can become a ham!

OPERATOR LICENSE CLASSES AND EXAM REQUIREMENTS

Today, there are three amateur operator licenses issued by the FCC – Technician, General, and Extra. Each license requires progressively higher levels of learning and proficiency, and each gives you additional operating privileges. This is known as *incentive licensing* – a method of strengthening the amateur service by offering more radio spectrum privileges in exchange for more operating and electronic knowledge.

There is no waiting time required to upgrade from one amateur license class to another, nor any required waiting time to retake a failed exam. You can even take all three examinations at one sitting if you're really brave! *Table 3-1* details the amateur service license structure and required examinations.

Table 3-1: Current Amateur License Classes and Exam Requirements
(Effective April 15, 2000)

License Class	Exam Element	Type of Examination
Technician Class	2	35-question, multiple-choice written examination. Minimum passing score is 26 questions answered correctly (74%).
General Class	3	35-question, multiple-choice written examination. Minimum passing score is 26 questions answered correctly (74%).
Extra Class	4	50-question, multiple-choice written examination. Minimum passing score is 37 questions answered correctly (74%).

ABOUT THE WRITTEN EXAMS

What is the focus of each of the written examinations, and how does it relate to gaining expanding amateur radio privileges as you move up the ladder toward your Extra Class license? *Table 3-2* summarizes the subjects covered in each written examination element.

Table 3-2. Question Element Subjects

Exam Element	License Class	Subjects
Element 2	Technician	Elementary operating procedures, radio regulations, and a smattering of beginning electronics. Emphasis is on VHF and UHF operating.
Element 3	General	HF (high-frequency) operating privileges, amateur practices, radio regulations, and a little more electronics. Emphasis is on HF bands.
Element 4	Extra	Basically a technical examination. Covers specialized operating procedures, more radio regulations, formulas and heavy math. Also covers the specifics on amateur testing procedures.

No Jumping Allowed

All written examinations for an amateur radio license are additive. You *cannot* skip over a license class or by-pass a required examination as you upgrade from Technician to General to Extra. For example, to obtain a General Class license, you must first take and pass the Element 2 written examination for the Technician Class license, plus the Element 3 written examination. To obtain the Extra Class license, you must first pass the Element 2 (Technician) and Element 3 (General) written examinations, and then successfully pass the Element 4 (Extra) written examination.

> *Still want to learn the code?*
> You can obtain Gordo's CW audio courses by visiting ***www.w5yi.org*** or by calling 800-669-9594.

TAKING THE ELEMENT 2 EXAM

Here's a summary of what you can expect when you go to the session to take the Element 2 written examination for your Technician Class license. Detailed information about how to find an exam session, what to expect at the session, what to bring to the session, and more, is included in Chapter 5.

All amateur radio examinations are administered by a team of at least three Volunteer Examiners (VEs) who have been accredited by a Volunteer Examiner Coordinator (VEC). The VEs are licensed hams who volunteer their time to help our hobby grow.

Examination sessions are organized under the auspices of an approved VEC. A list of VECs is located in the Appendix on page 197. The W5YI-VEC and the ARRL-VEC are the 2 largest examination groups in the country, and they test in all 50 states. Their 3-member, accredited examination teams are just about everywhere. So when you call the VEC, you can be assured they probably have an examination team only a few miles from where you are reading this book right now!

The Element 2 written exam is a multiple-choice format. The VEs will give you a test paper that contains the 35 questions and multiple choice answers, and an answer sheet for you to complete. Take your time! Make sure you read each question carefully and select the correct answer. Once you're finished, double check your work before handing in your test papers.

The VEs will score your test immediately, and you'll know before you leave the exam site whether you've passed. Chances are very good that, if you've studied hard, you'll get that passing grade!

> *Want to find a test site fast?*
> Visit the W5YI-VEC website at **www.w5yi.org**, or call 800-669-9594.

GETTING YOUR FIRST CALL SIGN

Once the VE team scores your test and you've passed, the process of getting your official FCC Amateur Radio License begins – usually that same day.

At the exam site, you will complete NCVEC Form 605, which is your application to the FCC for your license. If you pass the exam, the VE team will send on the required paperwork to their VEC. The VEC reviews the paperwork submitted by your exam team and then files your application with the FCC. This filing is done electronically, and your license will be granted and your call sign posted on the FCC's website within a few days. As soon as you see your new call sign, you are permitted to go on the air as a licensed amateur – even before your paper license arrives in the mail! Your first call sign is assigned by the FCC's computer, and you have no choice of letters. However, once you have that call sign, you can apply for a Vanity Call Sign. See Chapter 5 for more details.

HOW MANY CLASSES OF LICENSES?

Once you've passed your Element 2 exam and go on the air as a new Technician Class operator, you'll be talking to fellow hams throughout the U.S. and around the world. Here's a summary of the new and "grandfathered" licenses that your fellow amateurs may hold.

New License Classes

Following the FCC's restructuring of Amateur Radio that took effect April 15, 2000, there are just three written exams and three license classes – Technician, General, and Extra. But persons who hold licenses issued prior to April 15, 2000, may continue to hold onto their license class and continue to renew it every 10 years for as long as they wish.

"Grandfathered" Licensees

As mentioned previously, individuals licensed prior to April 15, 2000, will continue to enjoy band privileges based on their licenses. So, once you get on the air with your new Technician Class privileges, every now and then you might meet a Novice operator while yakking on 10 meters, or sending CW on 15, 40, or 80 meters.

And you'll see some older licenses saying "Technician Plus," which belong to grandfathered Technician Class operators licensed prior to April 15, 2000, who passed their 5-wpm code test and who get to keep their code credit indefinitely as long as they renew their license. Technician Plus operators will have their licenses renewed as Technician Class.

And then there are the Advanced Class operators who may continue to hold onto their license class designation until they finally decide to move up to Extra Class – without any further code requirement.

When you look at the Frequency Charts in this book that tell you the various band privileges, you will continue to see designated sub-bands for Extra, Advanced, General, Tech, and Novice. These sub-bands privileges haven't changed – any ham licensed prior to April 15, 2000, is automatically "grandfathered" to their original frequency privileges.

IT'S EASY!

Probably the primary pre-requisite for passing any amateur radio operator license exam is the will to do it. If you follow my suggestions in this book, your chances of passing the Technician exam are excellent.

Yes, indeed, the year 2000 brought some big changes to ham radio, and everyone comes out a winner! There has never been a better time to join the ranks of ham radio hobbyists. So study hard! We hope to hear you on the air very soon.

4

Getting Ready for the Exam

Your Technician Class written examination will consist of 35 multiple-choice questions taken from the 394 questions that make up the 2010-14 Element 2 pool. Each question on your examination and the multiple-choice answer will be identical to what is contained in this book.

This chapter contains the official, complete 394-question FCC Element 2 Technician Class question pool from which your examination will be taken. Again, your exam will contain 35 of these questions, and you must get 74% of the questions correct – which means you must answer 26 questions correctly in order to pass. This chapter also contains important information about how the exam is constructed using the questions taken from the Element 2 pool.

Your examination will be administered by a team of 3 or more Volunteer Examiners (VEs) – amateur radio operators who are accredited by a Volunteer Examiner Coordinator (VEC). You will receive a Certificate of Successful Completion of Examination (CSCE) when you pass the examination. This is official proof that you have passed the exam and it will be given to you before you leave the exam center. In just a few days, you'll find your call letters on the FCC website, and as soon as you know them you are licensed to go on the air!

THE 2010-14 QUESTION POOL

Development and maintenance of all three of the Amateur Radio Question Pools is the responsibility of the National Conference of Volunteer Examination Coordinators (NCVEC). This important job was assigned to the NCVEC by the Federal Communications Commission in 1986. The NCVEC appoints a Question Pool Committee from its membership to review and update the Technician, General, and Extra class question pools. Each pool is updated once every 4 years.

The exam questions in this 2010-14 Element 2 Technician Class Question Pool were authored by a team of five active amateur radio operators. Each of these active hams helped form this new pool by removing old questions on such topics as obsolete technology; adding new questions on such topics as the newest operating techniques and technology, and current rules; and, most important, selecting questions they feel a brand new ham radio operator needs to know to get on the air successfully!

Congratulations to the NCVEC Question Pool Committee members who spent many months working on this new Technician Class question pool: Roland Anders, K3RA, Laurel Amateur Radio Club VEC, Chairman of the QPC; and committee members Tom Fuszard, KF9PU, the Milwaukee Radio Amateurs Club VEC; Perry Green, WY1O, ARRL VEC; Larry Pollock, NB5X, W5YI-VEC; and Jim Wiley, KL7CC, Anchorage Amateur Radio Club VEC.

Thanks also go to many technical specialists, including Ward Silver, N0AX, a regular contributor to "QST" magazine, who contributed new questions and input to this new pool. I also was pleased to submit new questions and revisions to the QPC for this 2010-14 Element 2 Technician Class question pool.

WHAT THE EXAMINATION CONTAINS

The examination questions and the multiple-choice answers (one correct answer and three "distracters") for all license class levels are public information. They are widely published and are identical to those in this book. *There are no "secret" questions.* FCC rules prohibit any examiner or examination team from making any changes to any questions, including any numerical values. No numbers, words, letters, or punctuation marks can be altered from the published question pool. By studying all 394 Element 2 questions in this book, you will be reading the same exact questions that will appear on your 35-question Element 2 written examination. But which 35 out of the 394 total questions?

Table 4-1 shows how the Element 2, 35-question examination will be constructed. The question pool is divided into 10 subelements. Each subelement covers a different subject and is divided into topic areas. For example, on your Element 2 examination, of the 44 total pool questions in Subelement T8 on radio activities, 4 questions will be on your exam. Are you a little rusty on junior-high-school-level division and multiplication? Out of the 45 total pool questions on Ohm's Law and power calculations, only FOUR QUESTIONS that could deal with simple math will appear on your upcoming 35 Q & A exam!

Table 4-1. FCC Element 2 Technician Class Question Pool

Subelement	Topic	Total Questions	Exam Questions
T1	FCC rules, descriptions and definitions for the amateur radio service, operator and station license responsibilities	68	6
T2	Operating Procedures	31	3
T3	Radio wave characteristics, radio and electromagnetic properties, propagation modes	33	3
T4	Amateur radio practices and station setup	22	2
T5	Electrical principles, math for electronics, electronic principles, Ohm's Law	45	4
T6	Electrical components, semiconductors, circuit diagrams, component functions	47	4
T7	Station equipment, common transmitter and receiver problems, antenna measurements and troubleshooting, basic repair and testing	47	4
T8	Modulation modes, amateur satellite operation, operating activities, non-voice communications	44	4
T9	Antennas, feed lines	22	2
T0	AC power circuits, antenna installation, RF hazards	35	3
TOTALS		394	35

All Volunteer Examination teams use the same multiple-choice question pool. This uniformity in study material ensures common examinations throughout the country. Most exams are computer-generated, and the computer selects one question from each topic area within each subelement for your upcoming Element 2 exam.

Trust me – trust me, every question on your upcoming Element 2 exam will look very familiar to you by the time you finish studying this book

QUESTION CODING

Each and every question in the 396 Element 2, Technician Class question pool is numbered using a **code.** *The coded numbers and letters reveal important facts about each question!*

The numbering code always contains 5 alphanumeric characters to identify each question. Here's how to read the question number so you know exactly how the examination computer will select one question out of each group for your exam. Once you know this information, you can increase your odds of achieving a "max" score on the exam, especially if there is a specific group of questions which seems impossible for you to memorize or understand. When you get to Element 4, Extra Class – a very tough exam – this trick will really come in handy!

Let's pick a typical question out of the pool – T8A03 – and let me show you how this numbering code works:

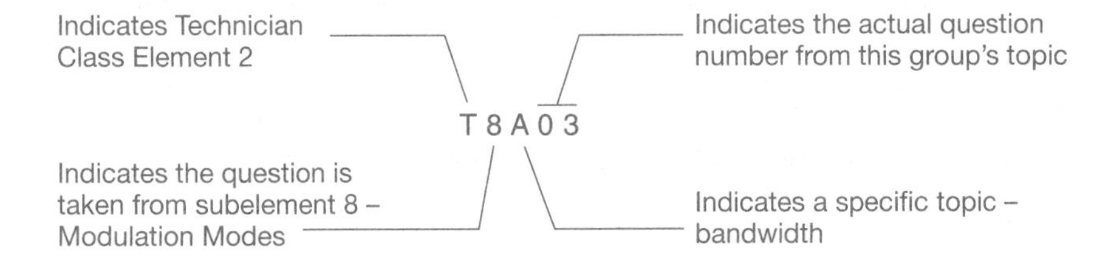

Figure 4-1. Examination Question Coding

- The first character "T" identifies the license class question pool from which the question is taken. "T" is for Technician. "G" would be for General, and "E" would be for Extra.
- The second digit, an "8", identifies the subelement number, 1 through 0. Technician subelement 8 deals with modulation modes, amateur satellite operations, operating activities, non-voice communications.
- The third character, "A", indicates the topic area within the subelement. Topic "A" deals with FCC modulation modes.
- The fourth and fifth digits indicate the actual question number within the subelement topic's group. The "03" indicates this is the third question about the use of single sideband modulation. There are 11 individual questions in topic area T8A, *but only one out of this topic group will appear on the test.*

Here's the Secret Study Hint

Only one exam question will be taken from any single group! A computer-generated test is set up to take one question from one single topic group. It cannot skip any one group, nor can it take any more than one question from that group.

Your upcoming Element 2, Technician Class, written exam is relatively easy with no bone-crusher math formulas. The same thing is true for the Element 3 exam, General Class. But when you get to the *Extra Class* book, there may be one or two groups that have formulas so complicated that you may want to wait until the very end to digest them. And if you decide to skip them completely, guess what – how many questions out of any one group? That's right, only one per group. This means you are not going to get hammered on any upcoming test with a whole bunch of questions dealing with a specific group item. Great secret, huh?

Study Time

The Technician Class question pool in this book is valid from July 1, 2010, through June 30, 2014. This pool contains 394 fresh, new questions written to a middle-school reading level. The Question Pool Committee enlisted outside educational experts to make sure kids, adults, and applicants with minor reading disabilities will comprehend the meaning of the very short-sentence question, and to be able to spot the correct multiple-choice answer – also very short – out of the 4 possible answers. These new questions are specifically designed to give the new Technician Class operator useful, practical knowledge so they can begin operating their equipment as soon as their call letters show up on the FCC database!

So how long will it take you to prepare for your upcoming exam? Probably about 30 days to work through this book and learn the material. Once you're ready, call The W5YI VEC at 800-669-9594 to find an exam location near you.

QUESTIONS REARRANGED FOR SMARTER LEARNING

The first thing you'll notice when you look at how the Element 2 question pool is presented in this book is that I have *completely rearranged the entire Technician Class question pool* to follow my weekend ham radio classes. This rearrangement will take you *logically* through each and every one of the 394 Q & A's. The questions are arranged into 20 topic areas in a way that eliminates the need for you to jump between topic groups or subelements to match up questions on similar topics.

For example, I have taken all of the questions in the pool that talk about where you can operate your ham radio and grouped them together into one area that allows you to better understand all of the material that relates to this topic. This arrangement of the Q&A's follows a natural learning process beginning with what ham radio is, what radio waves are, how they get from here to there, a peek under the cover of the modern Technician Class ham set, and ends with safety issues to keep you safe and sound in your new amateur radio hobby.

Trust me, the reorganization of all of the test questions in the 394 pool has been tested and finely-tuned in hundreds of my weekend classes. This method of learning WORKS! You will probably cut the amount of study time in half simply by following the question pool from front to back as presented here in my book!

Let me assure that each and every Element 2 Q&A is in this book. A cross-reference of all 394 questions is found on pages 205 to 206 in the back of the book, along with the syllabus used by the QPC to develop this new Element 2 pool.

This book – and my Technician Class audio course – contain all 394 Technician Class questions, 4 possible answers, the noted correct answer, and my upbeat description of how the correct answer works into the real world of amateur radio. We also include ***KEY WORDS*** that will help you remember the correct answer and provide you with a fast review of the entire question pool just before you sit for the big Tech exam. And be sure to look for our friend Elmer. He has many HAM HINTS to share with you, giving special tips and insights into ham radio operating to better illustrate specific Technician Class on-air techniques. At the end of each section of Q&As, we provide Website Resources – web addresses that can provide you with hours of fascinating study on selected "hot topics" that will help you understand the real world of ham radio behind the Q & A's. You can visit the site while you study, or visit them after you've earned your Technician Class licenses and are on the air.

When you visit some of these websites, it may not be immediately apparent why we are suggesting that you go there. Some addresses take you to the sites of local ham clubs. Most of these are specialty clubs, and they contain lots of information on how to operate on repeaters, or satellites, or provide educational resources on learning about electronics or antennas or how radios work.

While we worked hard to make sure all of these addresses are current at the time of publication, websites move, addresses change, or sites simply go away. So if you find an address that doesn't work, feel free to drop us an e-mail so we can update our book. Here's our e-mail address: masterpubl@aol.com.

How to Read the Questions

Using an actual question from page 31, here is a guide to explain what it is you will be studying as you go through all of the Q&As in the book:

Official Q&A

T1A02 What agency regulates and enforces the rules for the Amateur Radio Service in the United States?

A. FEMA.
B. The ITU.
C. The FCC.
D. Homeland Security.

The Federal Communications Commission (FCC) makes and enforces all Amateur Radio rules in the United States. [97.1] **ANSWER C**

Key Words to Remember

FCC Part 97 Rule Citation

Correct Answer

Topic Areas

Here is a list of my topic areas showing the page where it starts in the book. Again, there is a complete cross reference list of the Q&As in numerical order on pages 205 to 206 in the Appendix, along with the official Question Pool Syllabus.

STUDY SUGGESTIONS

Here are some suggestions to make your learning easier:

1. Read over each multiple-choice answer carefully. Some answers start out looking good, but turn bad during the last 2 or 3 words. If you speed read the answers, you could very easily go for a wrong answer. Also, while they won't change any words in the answers, they will sometimes scramble the A-B-C-D order.
2. Keep in mind that there is only one question on your test that will come from each group, and track how many groups in each sub-element.
3. Give this book to a friend, and ask him or her to read you the correct answer. You now give her the question wording.
4. Mark the heck out of your book! When the pages begin to fall out, you're probably ready for the exam! Take it with you everywhere you go.
5. My book is available on audio CDs, too. So if you'd like to listen to me read the questions and answers to you while you're driving your car, riding your bike, or laying on the beach, I can do that for you. The CD symbol, disk number, and track number at the beginning of each topic section keys this book to the audio book. You can get my book on audio CD where you purchased this book, or by calling 800-669-9594, or by visiting www.w5yi.com.

   CD 1 TRACK 2

6. Highlight the keywords one week before the test. Then speed read the brightly highlighted key words twice a day before the exam.

Are you ready to work through the 394 Q & A's? Put a check mark by the easy ones that you may already know the answer for, and put a little circle by any question that needs a little bit more study. Save your highlighting work until a few days before your upcoming test. Work the Q & A's for about 30 minutes at a time. I'll drop in a little bit of humor to keep you on track; and if you actually need my live words of encouragement, you can call me on the phone Monday through Thursday, 10:00 a.m. to 4:00 p.m. (California time), 714/549-5000.

THE QUESTION POOL, PLEASE

Okay, this is the big moment – your Technician Class, Element 2, question pool. Don't freak out and get overwhelmed with the prospect of learning 394 Q & A's. You will find that the topic content is repeated many times, so you're really going to breeze through this test without any problem!

TRACK 2

About Ham Radio

T1A01 For whom is the Amateur Radio Service intended?

A. Persons who have messages to broadcast to the public.
B. Persons who need communications for the activities of their immediate family members, relatives and friends.
C. Persons who need two-way communications for personal reasons.
D. Persons who are interested in radio technique solely with a personal aim and without pecuniary interest.

Amateur radio operators are affectionately called "hams." Every "ham" will give you a story on where that word came from, but we do our radio communicating as a hobby, and a service, NOT because we need a free radio service to talk with family and friends, nor do we become a ham to fulfill a specific two-way radio need. We don't broadcast to the general public, so our points of contact are to fellow ham radio operators to explore the fascinating world of radio propagation. We do this without receiving compensation (pecuniary interest). We are ***ham radio operators to satisfy our personal interest in radio techniques!*** [97.3(a)(4)] **ANSWER D**

There is no minimum age requirement for holding an FCC Amateur Radio License.

T1A02 What agency regulates and enforces the rules for the Amateur Radio Service in the United States?

A. FEMA.
B. The ITU.
C. The FCC.
D. Homeland Security.

The Federal Communications Commission (FCC) makes and enforces all Amateur Radio rules in the United States. [97.1] **ANSWER C**

T1C10 How soon may you operate a transmitter on an amateur service frequency after you pass the examination required for your first amateur radio license?

A. Immediately.
B. 30 days after the test date.
C. As soon as your name and call sign appear in the FCC's ULS database.
D. You must wait until you receive your license in the mail from the FCC.

Good news! You will not need to wait very long to begin operating with your first Amateur Radio license. ***As soon as you see your FCC license grant on the Internet***, or make a phone call to someone who can read you your new call sign from the FCC database, *you are on the air!* The paper copy of your license will come in about three to four weeks. If your Volunteer Examination Team and Coordinator file electronically, you should see a license grant within 48 hours of electronic submission. [97.5(a)] **ANSWER C.**

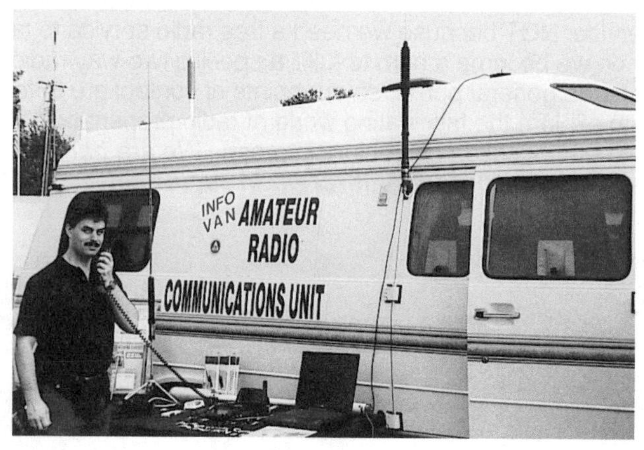

Make sure the FCC has issued your call sign before you go on the air for the first time.

Ham Hint: *The first thing you'll receive at your mailing address is* your new license! *Even though you might already be on the air because you spot your new call sign on the FCC ULS database, make sure you receive the paper copy of the license at your address as a double-check that your postal address is indeed accurate. Undeliverable mail from the FCC could lead to serious problems. Make sure your address is absolutely correct! And if you move, make absolutely sure you notify the FCC immediately with your new address.*

T1C08 What is the normal term for an FCC-issued primary station/operator license grant?

A. Five years.
B. Life.
C. Ten years.
D. Twenty years.

Amateur station licenses are granted for a ***term of 10 years***. You will need to renew your license at that time, and renewing will probably be done as an on-line ULS application to the FCC. There is no exam required to renew a valid ham radio license. [97.25] **ANSWER C.**

T1C09 What is the grace period following the expiration of an amateur license within which the license may be renewed?

A. Two years.
B. Three years.
C. Five years.
D. Ten years.

You are not allowed to operate during the 2-year grace period; however, ***you keep your privileges for 2 years***. After that, they are lost for good. So is your call sign. Don't forget to renew! [97.21(a)(b)] **ANSWER A.**

UNITED STATES OF AMERICA
FEDERAL COMMUNICATIONS COMMISSION

AMATEUR RADIO LICENSE

KB9SMG

T1C11 If your license has expired and is still within the allowable grace period, may you continue to operate a transmitter on amateur service frequencies?

A. No, transmitting is not allowed until the ULS database shows that the license has been renewed.
B. Yes, but only if you identify using the suffix "GP."
C. Yes, but only during authorized nets.
D. Yes, for up to two years .

How could an active, licensed amateur radio operator let their license expire? They move. You must inform the FCC when your mailing address changes. Here's why–VEC groups may send a reminder letter that your license needs to be renewed. You may do it yourself on the computer or, for a small fee, the VEC system can help you. What? You didn't get the reminder letter? I can tell you why. You moved! FCC mail is never forwarded, and I can tell you (not first hand) when they get returned mail, they have the option of canceling your license. Sometimes, another ham operator will look up your call sign on a database, and let you know you are operating in violation, within your 2 year grace period. Remember, ***no operating is allowed as soon as your license lapses***. So keep your address current with the FCC! [97.21(b)] **ANSWER A.**

Ham Hint: *When you take your upcoming examination for a new license, it will be fellow ham radio operators, accredited by a VEC, who will give you the exam. It takes three or more accredited Volunteer Examiners to conduct a test session. Each of these examiners also will be wearing a plastic accreditation ID card so there will be no mistaking who's who in the examination room. These 3 examiners volunteer their time to conduct the test sessions, so go overboard in thanking each of them with a handshake for offering you your first license exam. Say hi from "Gordo," too!*

T1A10 What is the FCC Part 97 definition of an amateur station?

A. A station in an Amateur Radio Service consisting of the apparatus necessary for carrying on radio communications.
B. A building where Amateur Radio receivers, transmitters, and RF power amplifiers are installed.
C. Any radio station operated by a non-professional.
D. Any radio station for hobby use.

Your ham station will consist of ***a radio transmitting device*** that will be ***used for amateur communications***. This station can go anywhere that you go, and might be mobile, a base, or a handheld set. You also are allowed to choose any authorized frequency on any authorized band on which you have Technician Class operating privileges. You can change radio equipment type at anytime. [97.3(a)(5)]
ANSWER A.

Read the Rules - Heed the Rules

Here are some links to internet websites and e-mail addresses where you can find out more about the FCC's Amateur Radio rules and regs.

www.gpoaccess.gove/ecfr/ [title 47] then [Part 97].

Also see Parts 0, 1, 2, 17,and 214.

http://wireless.fcc.gov/ [amateur] or [ULS]

Read "The Rules Say" at WorldRadio Online at www.cq-amateur-radio.com

Questions about the amateur service rules? Be informed!
www.w3beinformed.org

or e-mail your questions to john@johnston.net

FCC enforcement reports are at: www.fcc.gov/eb/AmateurActions/

Report rule violations to: fccham@fcc.gob

Call Signs

T1F03 When is an amateur station required to transmit its assigned call sign?

A. At the beginning of each contact, and every 10 minutes thereafter.
B. At least once during each transmission.
C. At least every 15 minutes during and at the end of a contact.
D. At least every 10 minutes during and at the end of a contact.

Give your call letters regularly – ***every 10 minutes*** and at the end of your transmission. Remember, even though the law doesn't require that you give them at the beginning of the transmission, it makes good sense to start out with your call letters. [97.119(a)] **ANSWER D.**

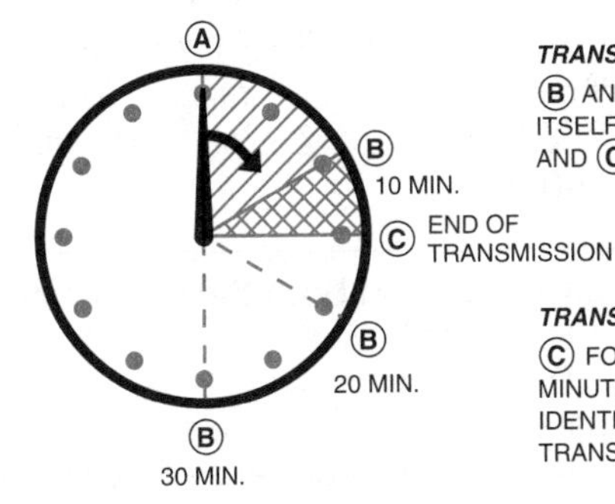

Identifying Amateur Transmissions

T1C02 Which of the following is a valid US amateur radio station call sign?

A. KMA3505.
B. W3ABC.
C. KDKA.
D. 11Q1176.

Ham radio call signs, for the United States, begin with A, K, N, or W. They also have a single number 0 through 9. Answer A has too many letters and numbers for a ham call sign. Looks like police radio to me. Answer C is incorrect because there are no numbers, all letters, and is the call sign of the very first commercial broadcast radio station in the US, located in Pittsburgh. And Answer D reminds me of my old CB call sign, 11W1769. That leaves us with only Answer B as the correct format for a US ham radio call sign. **ANSWER B.**

Many hams order license plates with their call sign.

T2B09 Which of the following methods is encouraged by the FCC when identifying your station when using phone?

A. Use of a phonetic alphabet.
B. Send your call sign in CW as well as voice.
C. Repeat your call sign 3 times.
D. Increase your signal to full power when identifying.

When you get your new call sign, hardly anyone else on the repeater will be familiar with it, and you should ***use the International Phonetic Alphabet to make your individual letters recognized*** by substituting a word for each letter. Memorize the phonetic alphabet, and use it often. When you say your call sign phonetically, and indicate you are listening for a response, put some zing into it! This is like fishing – sound excited on the airwaves and you'll get some excitement coming back to your first call. Don't be deadpan – have some fun in your voice and you'll get lots of return calls. [97.119(b)(2)] **ANSWER A.**

ITU International Phonetic Alphabet

A – Alfa – AL fah	J – Juliet – JEW lee ett	S – Sierra – SEE air rah
B – Bravo – BRAH voh	K – Kilo – KEY loh	T – Tango – TANG go
C – Charlie – CHAR lee	L – Lima – LEE mah	U – Uniform – YOU nee form
D – Delta – DELL tah	M – Mike – MIKE	V – Victor – VICK tah
E – Echo – ECK ohh	N – November – NO vem ber	W – Whiskey – WISS key
F – Foxtrot – FOKS trot	O – Oscar – OSS car	X – X-ray – ECKS ray
G – Golf – GOLF	P – Papa – PAH pah	Y – Yankee – YANG kee
H – Hotel – HOH tel	Q – Quebec – KEH beck	Z – Zulu – ZOO loo
I – India – IN dee ah	R – Romeo – ROW me o	

T1C01 Which type of call sign has a single letter in both the prefix and suffix?

A. Vanity.
B. Sequential.
C. Special event.
D. In-memoriam.

"Calling CQ from W6V, Veterans' Memorial Station." Wow, nice short call sign, huh? W 6 V? This is a ***special event call sign***, and special event call signs are granted only to those stations that may be operating on a special day (Memorial Day weekend), offering other hams the ability to exchange communications on that special day, at a special facility. [97.3(a)(11)(iii)] **ANSWER C.**

T8C06 For what purpose is a temporary "1 by 1" format (letter-number-letter) call sign assigned?

A. To honor a deceased relative who was a radio amateur.
B. To designate an experimental station.
C. For operations in conjunction with an activity of special significance to the amateur community.
D. All of these choices are correct.

Every year the Veterans Administration Center in our local area comes on the air with their ***special event call sign***, W6V. They may only use this call sign for a limited time to mark a special activity or their special station during a specific time of year, such as Veterans Day. **ANSWER C.**

T1F12 How many persons are required to be members of a club for a club station license to be issued by the FCC?

A. At least 5.
B. At least 4.
C. A trustee and 2 officers.
D. At least 2.

The FCC closely monitors club station applications for a club station call sign. It requires ***at least 4 persons to make up the club***, and the club must have a name, a document of organization, a list of management, and the primary purpose devoted to amateur radio service activities. The FCC will not tolerate a ham operator "collecting" multiple different club call signs just for the fun of it. [97.5(b)(2)]
ANSWER B.

Ham Hint: *The FCC enforcement division regularly sends correspondence to applicants requesting more than one club call sign asking them to justify the request, especially when the additional "clubs" are nearly identical in nature, except for a one or two word difference in their names.*

Gordo's Ham Club — *The Official Club of Gordo*

Ham Clubs for Gordo — *The Fraternity Club of Gordo Hams*

Submitting an application for a fictitious club could lead to license revocation! Many an Extra Class ham had to "give back" 14 out of 15 assigned club call signs. To learn how your ham radio club might obtain its own club call sign, contact the Club Call Sign Administrator at the W5YI-VEC, www.w5yi.org or 800-669-9594. Be prepared to justify your application for a club call sign.

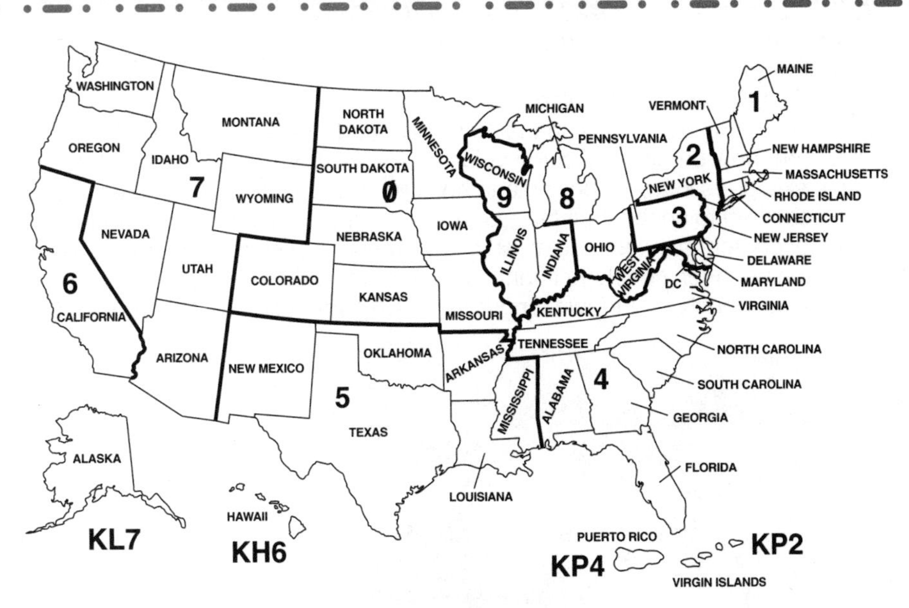

U.S. Call Sign Areas

The number in your new call sign is determined by your permanent mailing address. If you're in Vermont, your call sign might be KB1DOG. In Texas? You're in "5-land." Wisconsin is in call sign district 9. When you move, you may still keep your call sign.

T1F01 What type of identification is being used when identifying a station on the air as "Race Headquarters"?

A. Tactical call.
B. Self-assigned designator.
C. SSID.
D. Broadcast station.

Tactical call signs are permitted on the ham bands to identify a specific aspect of your operation. If you are running sled dog races in Fairbanks, other operators may simply call "Race Headquarters" if you are the station at race headquarters. However, every 10 minutes, be sure to give your own call sign, as well as the ***tactical call*** name you have been using. **ANSWER A.**

T1F02 When using tactical identifiers, how often must your station transmit the station's FCC-assigned call sign?

A. Never, the tactical call is sufficient.
B. Once during every hour.
C. Every ten minutes.
D. At the end of every communication.

You will be the only one in the World with your unique amateur radio call sign. You can likely get your ham radio call sign on your vehicle license plates, too! And when operating on the air, it's always a good idea to give your call sign phonetically. When operating a ham radio station at a big event, you are permitted to use tactical call signs, such as "Checkpoint Charlie." However, the rules still require that you ***give your own station call sign every ten minutes*** along with your tactical call sign. So every ten minutes, if you have been active on the air, give your own call sign. [97.119(a)] **ANSWER C.**

T1F04 Which of the following is an acceptable language for use for station identification when operating in a phone sub-band?

A. Any language recognized by the United Nations.
B. Any language recognized by the ITU.
C. The English language.
D. English, French, or Spanish.

Today is an exciting adventure on the 10 meter band. As a Technician Class operator, your new privileges include voice emissions from 28.300 to 28.500 MHz. When the skip goes long, it's possible to hook up with a station in a foreign country. You are permitted to speak their language as a courtesy to the other operator. However, every ten minutes, you must give your own call sign in English. Even though you have been speaking fluent Italian for the last 9 minutes and 59 seconds, its now time to identify in our own language — ***English***. [97.119(b)] **ANSWER C.**

Testing your radio? Give your call sign! In English!

T1F06 Which of the following formats of a self-assigned indicator is acceptable when identifying using a phone transmission?

A. KL7CC stroke W3.
B. KL7CC slant W3.
C. KL7CC slash W3.
D. All of these choices are correct.

Your Alaska station is now operating mobile in Maryland, the 3rd call sign district. So local operators know you are in the east coast area, you would use your regular call sign followed by the word stroke, or the word slant, or digital indicator of forward slash mark and then "W3". ***All of these choices are correct***. This way, locals won't think Alaska propagation is coming in on the local 2 meter band! They know you are in their neighborhood. [97.119(c)] **ANSWER D.**

T1F07 Which of the following restrictions apply when appending a self-assigned call sign indicator?

A. It must be more than three letters and less than five letters.
B. It must be less than five letters.
C. It must start with the letters AA through AL, K, N, or W and be not less than two characters or more than five characters in length.
D. It must not conflict with any other indicator specified by the FCC rules or with any call sign prefix assigned to another country.

Today we have you running a public service event at a bicycle rally medical aid station. There are 3 medical aid stations on the course, and your station is "Mash One." Since "Mash One" doesn't sound like any FCC or other country call sign, you would be okay to use it as a location identifier, remembering to always use your official FCC-issued call sign every 10 minutes. ***Tactical call signs ARE permitted as long as they don't sound like a US or foreign call sign***. [97.119(c)] **ANSWER D.**

T1B01 What is the ITU?

A. An agency of the United States Department of Telecommunications Management.
B. A United Nations agency for information and communication technology issues.
C. An independent frequency coordination agency.
D. A department of the FCC.

The United States, like most other countries, is a member of the ***International Telecommunications Union. The ITU is a worldwide United Nations agency*** that helps establish uniform and agreed-upon regulations and frequency rules for radio services of member nations. This includes amateur radio services of member nations, which all share similar radio frequency bands. Our own FCC has the big job of insuring that our radio services – commercial as well as amateur – are in compliance with ITU rules. [97.2(a)(28)] **ANSWER B.**

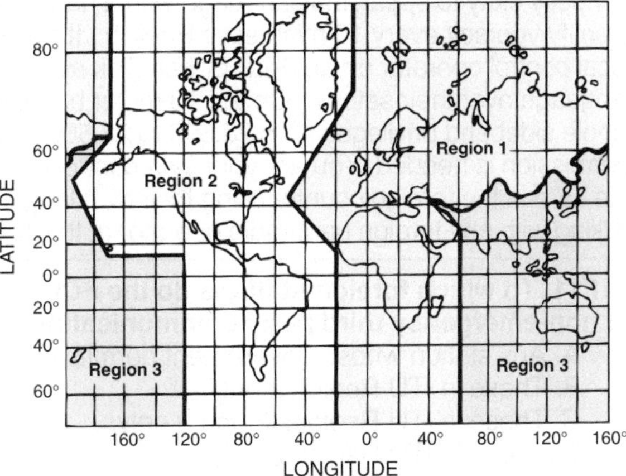

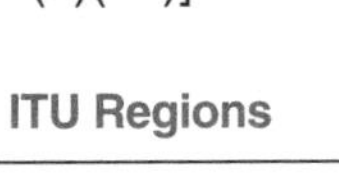
ITU Regions

T1B02 North American stations are located in which ITU region?

A. Region 1.
B. Region 2.
C. Region 3.
D. Region 4.

If you look at the ITU map, you will see that the United States is located in ITU Region 2. Even if we sail out to Hawaii, we are still in Region 2. But once you leave Hawaii, en route to the South Seas, guess what – you enter a new region! But for now we are in ***Region 2*** when plying the coast lines of the United States. **ANSWER B.**

T1C03 What types of international communications are permitted by an FCC-licensed amateur station?

A. Communications incidental to the purposes of the amateur service and remarks of a personal character.
B. Communications incidental to conducting business or remarks of a personal nature.
C. Only communications incidental to contest exchanges, all other communications are prohibited.
D. Any communications that would be permitted on an international broadcast station.

Ham radio operators are not allowed to conduct their own business over the air. Ham radio only for contesting is not what a good ham operator is about. We also don't transmit international broadcast messages to shortwave listeners. Our ham transmissions are of a ***personal nature, and the purpose of our communications*** is found in the amateur service Rules and Regulations, Part 97. [97.117] **ANSWER A.**

T1D01 With which countries are FCC-licensed amateur stations prohibited from exchanging communications?

A. Any country whose administration has notified the ITU that it objects to such communications.
B. Any country whose administration has notified the United Nations that it objects to such communications.
C. Any country engaged in hostilities with another country.
D. Any country in violation of the War Powers Act of 1934.

Your new Technician Class license gives you many opportunities to talk around the world. A few hams own the distinction of WORKED ALL COUNTRIES – a little more than 350 throughout the world. Currently, no ***foreign government***, nor our own, ***prohibits ham-to-ham communications***. If you speak a foreign language, it's perfectly okay to speak that language with a ham in that country. Just be sure to identify yourself every 10 minutes in English. It's also a good idea to check with the local control operator on an IRLP or Echolink system for permission to speak the other language over their setups. On the 10 meter band, where you have worldwide voice single sideband privileges from 28.300 to 28.500 MHz, no other control operator permission is needed. You are your own control operator, and you may speak with the ham in the foreign country using their own language. We are not prohibited from talking with any foreign ham radio operator at this time. [97.111(a)(1)] **ANSWER A.**

T1F11 To which foreign stations do the FCC rules authorize the transmission of non-emergency third party communications?

A. Any station whose government permits such communications.
B. Those in ITU Region 2 only.
C. Those in ITU Regions 2 and 3 only.
D. Those in ITU Region 3 only.

Third party agreements allow you to let a non-licensed friend speak over your ham radio set. This is fine with all other US hams here, but is ***only permitted when our government has a third party agreement*** in place with the other country's government. [97.115(a)] **ANSWER A.**

List of Countries Permitting Third-Party Traffic

Country	Call Sign Prefix	Country	Call Sign Prefix	Country	Call Sign Prefix
Antigua and Barbuda	V2	El Salvador	YS	Paraguay	ZP
Argentina	LU	The Gambia	C5	Peru	OA
Australia	VK	Ghana	9G	Philippines	DU
Austria, Vienna	4U1VIC	Grenada	J3	Pitcairn Island	VR6
Belize	V3	Guatemala	TG	St. Christopher & Nevis	V4
Bolivia	CP	Guyana	8R	St. Lucia	J6
Bosnia-Herzegovina	T9	Haiti	HH	St. Vincent & Grenadines	J8
Brazil	PY	Honduras	HR	Sierra Leone	9L
Canada	VE, VO, VY	Israel	4X	South Africa	ZS
Chile	CE	Jamaica	6Y	Swaziland	3D6
Colombia	HK	Jordan	JY	Trinidad and Tobago	9Y
Comoros	D6	Liberia	EL	Turkey	TA
Costa Rica	TI	Marshall Is	V6	United Kingdom	GB
Cuba	CO	Mexico	XE	Uruguay	CX
Dominica	J7	Micronesia	V6	Venezuela	YV
Dominican Republic	HI	Nicaragua	YN	ITU-Geneva	4U1ITU
Ecuador	HC	Panama	HP	VIC-Vienna	4U1VIC

T1C06 From which of the following may an FCC-licensed amateur station transmit, in addition to places where the FCC regulates communications?

A. From within any country that belongs to the International Telecommunications Union.
B. From within any country that is a member of the United Nations.
C. From anywhere within in ITU Regions 2 and 3.
D. From any vessel or craft located in international waters and documented or registered in the United States

When you get your new ham license, you are permitted to transmit anywhere our Federal Communications Commission regulates radio traffic, such as throughout the United States, and in our territories. We do have agreements in place for permission to operate in several other countries, and will learn more about this in up-coming pages of this book. But what happens when you are out sailing on the high seas, in international waters? If that ***sailboat is registered*** or ***documented in the US***, seen flying our stars and stripes on the stern flag staff, then you are good to go! You still must ***abide by the rules*** for your grade of license, and all the FCC rules are still in force, even though you may be 1000 miles out at sea. Be sure to get the permission of the ship's Captain (on a cruise ship, for instance) before going on the air with your little ham radio set. [97.5(a)(2)] **ANSWER D.**

T1C04 When are you allowed to operate your amateur station in a foreign country?

A. When the foreign country authorizes it.
B. When there is a mutual agreement allowing third party communications.
C. When authorization permits amateur communications in a foreign language.
D. When you are communicating with non-licensed individuals in another country.

Over 75 countries hold reciprocal operating agreements with the US. The permit to operate in different countries usually is obtained ahead of time, and some European countries don't require any special paperwork other than having copies of an Extra Class ***US license and reciprocal agreement*** paperwork. [97.107] **ANSWER A.**

The following CEPT countries allow U.S. Extra Class Amateurs to operate in their countries without a reciprocal license. Be sure to carry a copy of your FCC license and FCC Public Notice DA99-1098.

Austria	Finland	Liechtenstein	Slovenia
Belgium	France & its possessions	Lithuania	Spain
Bosnia & Herzegovina	Germany	Luxembourg	Sweden
Bulgaria	Greenland	Monaco	Switzerland
Croatia	Hungary	Netherlands	Turkey
Cyprus	Iceland	Netherlands Antilles	United Kingdom & its possessions
Czech Republic	Ireland	Norway	
Denmark	Italy	Portugal	
Estonia	Latvia	Romania	
Faroe Islands		Slovak Republic	

Countries Holding U.S. Reciprocal Agreements

Antigua, Barbuda	Chile	Greece	Liberia	Seychelles
Argentina	Colombia	Greenland	Luxembourg	Sierra Leone
Australia	Costa Rica	Grenada	Macedonia	Solomon Islands
Austria	Croatia	Guatemala	Marshall Is.	South Africa
Bahamas	Cyprus	Guyana	Mexico	Spain
Barbados	Denmark	Haiti	Micronesia	St. Lucia
Belgium	Dominica	Honduras	Monaco	St. Vincent and Grenadines
Belize	Dominican Rep.	Iceland	Netherlands	Surinam
Bolivia	Ecuador	India	Netherlands Ant.	Sweden
Bosnia-Herzegovina	El Salvador	Indonesia	New Zealand	Switzerland
Botswana	Fiji	Ireland	Nicaragua	Thailand
Brazil	Finland	Israel	Norway	Trinidad, Tobago
Canada[1]	France[2]	Italy	Panama	Turkey
	Germany	Jamaica	Paraguay	Tuvalu
		Japan	Papua New Guinea	United Kingdom[3]
		Jordan	Peru	Uruguay
		Kiribati	Philippines	Venezuela
		Kuwait	Portugal	

1. Do not need reciprocal permit
2. Includes all French Territories
3. Includes all British Territories

T1D02 On which of the following occasions may an FCC-licensed amateur station exchange messages with a U.S. military station?

A. During an Armed Forces Day Communications Test.
B. During a Memorial Day Celebration.
C. During a Independence Day celebration.
D. During a propagation test.

Once a year, during the ***Armed Forces Day Communications Test*** weekend drill, we get special permission to talk with Uncle Sam's finest, with them using their own military call signs. We transmit within our own ham band limits, with our own call signs, but with a receiver set to military frequencies outside of our ham bands, we listen on a different frequency, out of band! "Calling Coast Guard station November Foxtrot 1 1 4 Alpha Golf, this is amateur station WB6NOA, over." Perfectly legal during this 1 or 2 day event, once a year. [97.111(a)(5)] **ANSWER A.**

▾ IF YOU'RE LOOKING FOR	▾ THEN VISIT
Look up a Pal's Callsign	www.hamcall.net
Click on WB6NOA	www.QRZ.com
Ham Party Line!	www.eham.net
FCC Site with lots of Info	wireless.fcc.gov/
ARRL's Callsign Lookup	www.arrl.org/fcc/
More Call Lookups	www.arrl.org/FCC/FCCLook.PHP3
Available Vanity Calls	www.vanityhq.com

TRACK 3

Control

T1E01 When must an amateur station have a control operator?

A. Only when the station is transmitting.
B. Only when the station is being locally controlled.
C. Only when the station is being remotely controlled.
D. Only when the station is being automatically controlled.

Every ham station must have a responsible ***control operator when transmitting***. The FCC defines a control operator as: "An amateur operator designated by the licensee of an amateur station to also be responsible for the emissions from that station." If you are operating your ham equipment all by yourself, pedaling your bicycle along the river bank, you are both a control operator and station licensee, all rolled up into one! For the rest of your equipment, back at the house, no control operator is required providing that gear will not, cannot, and must not ever transmit without you (licensee and control op) being present at the home control point. [97.7(a)] **ANSWER A.**

When you operate your station you are the "control operator," and you are at the station's "control point."

T1E02 Who is eligible to be the control operator of an amateur station?

A. Only a person holding an amateur service license from any country that belongs to the United Nations.
B. Only a citizen of the United States.
C. Only a person over the age of 18.
D. Only a person for whom an amateur operator/primary station license grant appears in the FCC database or who is authorized for alien reciprocal operation.

There is no age limit or citizenship requirement to become a licensed amateur operator in the US. Your name appears on ***your ham license***, and you are eligible to be the ***control operator*** of that station, and, if given permission, any other ham station. Answer D is correct, but don't let all of the extra verbiage regarding reciprocal operation throw you off – you are good to go as a control operator. [97.7(a)] **ANSWER D.**

T8B01 Who may be the control operator of a station communicating through an amateur satellite or space station?

A. Only an Amateur Extra Class operator.
B. A General Class licensee or higher licensee who has a satellite operator certification.
C. Only an Amateur Extra Class operator who is also an AMSAT member.
D. Any amateur whose license privileges allow them to transmit on the satellite uplink frequency.

Any amateur can talk over the satellites. When you pass your new Technician Class exam, you'll be allowed to transmit through satellites on the satellite uplink frequency within your Technician Class privileges. You'll instantly join the mainstream of satellite activity! **ANSWER D.**

T1E03 Who must designate the station control operator?

A. The station licensee.
B. The FCC.
C. The frequency coordinator.
D. The ITU.

The ***control operator*** is the station licensee or another amateur radio licensee ***designated by the station licensee*** to be the control operator. It is up to the station licensee to either control his own station, or select an alternate control operator who is appropriately licensed. [97.103(b)] **ANSWER A.**

T1E07 When the control operator is not the station licensee, who is responsible for the proper operation of the station?

A. All licensed amateurs who are present at the operation.
B. Only the station licensee.
C. Only the control operator.
D. The control operator and the station licensee are equally responsible .

Your Extra Class Elmer comes over to your station and asks to use your 10 meter gear up on 29.6 MHz, FM, in an area that is beyond your Technician Class band privileges. This is okay, but your pal must give his own call sign, along with your call sign, and ***both of you are responsible for the transmissions*** since the gear is yours. Both of you are responsible! [97.103(a)] **ANSWER D.**

T1E04 What determines the transmitting privileges of an amateur station?

A. The frequency authorized by the frequency coordinator.
B. The class of operator license held by the station licensee.
C. The highest class of operator license held by anyone on the premises.
D. The class of operator license held by the control operator.

Soon you will be a Technician Class operator. Good for you! Did you know you have high frequency voice privileges on the 10 meter band, 28.3 to 28.5 MHz? Your friend who holds an Extra Class license enters the station and wants to run your station on 29.6 MHz FM. If you designate your friend as the control operator of your station, ***his or her privileges*** with the Extra Class ticket would allow operation well above your normal Technician Class band limits. [97.103(b)] **ANSWER D.**

When you operate from another ham's station, you use your license class privileges.

T1F08 When may a Technician Class operator be the control operator of a station operating in an exclusive Amateur Extra Class operator segment of the amateur bands?

A. Never.

B. On Armed Forces Day.

C. As part of a multi-operator contest team.

D. When using a club station whose trustee is an Amateur Extra Class operator.

As a Technician Class operator, you get some exciting worldwide high frequency bands down on 10 meters, 15 meters for CW, and 40 meters for CW, and 80 meters for CW. You really love Morse code, so could you tune down to the Extra Class portion of the band and run with the experts? ***Never – you must stay within your Technician Class band privileges***. [97.119(e)] **ANSWER A.**

T1E05 What is an amateur station control point?

A. The location of the station's transmitting antenna.

B. The location of the station transmitting apparatus.

C. The location at which the control operator function is performed.

D. The mailing address of the station licensee.

This is ***the spot where you have complete capabilities*** to turn the equipment on, or shut it off in case of a malfunction. Every ham radio station is required to have a control point. [97.3(a)(14)] **ANSWER C.**

The control point is the spot where you have complete capability to turn your equipment on or off. Today's mobil rigs put the keypad on the microphone, making it the control point.

T1E09 What type of control is being used when transmitting using a handheld radio?

A. Radio control.

B. Unattended control.

C. Automatic control.

D. Local control.

Whether you are talking over your little handheld, a mobile dual-band radio, or from a big powerful kilowatt home 6-meter station, if ***you are at the controls*** of the equipment this is officially called ***local control***. The equipment is the control point and – you guessed it – you're the control operator. [97.109(a)] **ANSWER D.**

T1E06 Under which of the following types of control is it permissible for the control operator to be at a location other than the control point?

A. Local control.

B. Automatic control.

C. Remote control.

D. Indirect control.

It has been snowing for 5 days, and the mountaintop repeater is functioning well. It is YOUR system, coordinated by a local frequency coordination council. Does that mean you must be at the actual repeater location, rather than controlling it from your warm basement radio room? Since the repeater is classified under ***"automatic control"*** you do not need to be up at that very chilly repeater site, nor at your gear in the basement. [97.109(d)] **ANSWER B.**

T1E10 What type of control is used when the control operator is not at the station location but can indirectly manipulate the operating adjustments of a station?

A. Local.
B. Remote.
C. Automatic.
D. Unattended.

"Anyone on the repeater want to talk to Antarctica? Here they come!" announced the repeater control operator. He's tying into IRLP (Internet Radio Linking Project) using the touch-tone pad on his handheld to activate the distant repeater on the Antarctica node using ***remote control***. [97.3] **ANSWER B.**

T1E08 What type of control is being used for a repeater when the control operator is not present at a control point?

A. Local control.
B. Remote control.
C. Automatic control.
D. Unattended.

About the only time you will see a big rack of ham radio equipment with no actual operator on duty is the ***automatically controlled*** station. This could be a repeater up on a mountain top, or a radio link atop a high rise. The law requires that all of the details indicating who is in charge be posted on the equipment in case of a malfunction during automatic control. [97.3(a)] **ANSWER C.**

T1E11 Who does the FCC presume to be the control operator of an amateur station, unless documentation to the contrary is in the station records?

A. The station custodian.
B. The third party participant.
C. The person operating the station equipment.
D. The station licensee.

Any time I let another licensee run ***my radio gear***, I always make a log entry to note who the control operator was at the time. If the FCC is monitoring, they will assume that ***I was the control operator***, unless I can provide documentation to the contrary in my station records. Always keep a log book, especially when other hams are using your gear or when passing third party traffic. [97.103(a)] **ANSWER D.**

T1D08 When may the control operator of an amateur station receive compensation for operating the station?

A. When engaging in communications on behalf of their employer.
B. When the communication is incidental to classroom instruction at an educational institution.
C. When re-broadcasting weather alerts during a RACES net.
D. When notifying other amateur operators of the availability for sale or trade of apparatus.

No transmitting messages on behalf of your employer, and no transmitting that your prize poodle just had a litter with each puppy going for $999. We may not be compensated by any local weather service to take weather reports far out at sea. In other words, no pay for transmitting over ham radio. There is an ***exception for ham radio operators who are school teachers***. They can be "on the clock" during classroom hours demonstrating ham radio for their students to learn what ham radio is all about. [97.113] **ANSWER B.**

School teachers can recieve their regular pay when teaching about ham radio.

T1F10 Who is accountable should a repeater inadvertently retransmit communications that violate the FCC rules?

A. The control operator of the originating station.
B. The control operator of the repeater.
C. The owner of the repeater.
D. Both the originating station and the repeater owner.

This question deals with the ham who inadvertently goes into a brain-fade while operating on a repeater. You are late to work, and talking on the air lamenting the terrific traffic congestion. A fellow worker ham says he's already in the parking lot and will tell the troops you are going to be late. Without thinking, you ask him to tell your secretary to postpone the sales meeting until tomorrow. *ERROR* – conducting a conversation over the air that applies to your direct business, as the sales manager for that company. ***You're responsible.*** [97.205(g)] **ANSWER A.**

Ham Hint: *Sometimes the repeater control operator, also stuck in the same traffic mess, will cheerfully say to give him a phone call when you get home. Or maybe he tells you, on the air (never recommended), that you violated the FCC rules. You reply that you hadn't had enough coffee yet, and you will never do it again.*

Rarely will the FCC get involved, and the likelihood of getting a Notice of Violation will be remote. Where the FCC DOES get involved is when a repeater control operator exhausts all on-air and telephone reasoning with a ham operator making rule violations over that control operator's repeater. The FCC will send a stern warning letter and demand a reply within 10 days. Never ignore a Notice of Violation from the FCC, nor ignore a letter from the FCC enforcement division. Shrugging off FCC correspondence will likely lead to license revocation. If you foul up on a repeater continuously, you are accountable for your own transmissions.

Website Resources

IF YOU'RE LOOKING FOR	THEN VISIT
Main Link to FCC Info	wireless.fcc.gov/rules.html
News for Hams Needing Special Help	www.handiham.org
All about amateur TV	www.atv-tv.org/atv-at.php
Microwave experimenters TV system	www.qsl.net/kc6ccc
N6HOG's weather web	http://n6hog.com
San Bernardino Microwave Society	www.ham-radio.com/sbms
ATV logger page	http://dxworld.com/atvlog.html
70 cm ATV info	www.downeastmicrowave.com
Request a free subscription to the ATV newsletter	atv-newsletter@hotmail.com

Mind The Rules

T1A03 Which part section of the FCC rules contains the rules and regulations governing the Amateur Radio Service?

A. Part 73.
B. Part 95.
C. Part 90.
D. Part 97.

Ham radio rules are found in Part 97. Part 97 falls under Title 47-Telecommunication. To eliminate the need for a wheelbarrow to haul around a huge book called Code of Federal Regulations, we separate out Title 47, Part 80 to End. It is a mere 4 inches thick, and includes the following radio services:

Part 80 Marine Radio
Part 87 Aviation Radio
Part 90 Land Mobile Radio
Part 95 Personal Radio, like CB and FRS rules
Part 97 Amateur Radio Service — THAT'S US
Part 101 Fixed Microwave Services

How about just the ***Part 97*** Rules and Regs in a smaller, handier format? Call 1-800-669-9594, and tell them Gordo sent you for an up-to-date Part 97 Rule book! **ANSWER D.**

Part 97 Amateur Radio regulation are contained in Title 47– Telecommunication. The full volume is 3 inches thick!.

T1D06 Which of the following types of transmissions are prohibited?

A. Transmissions that contain obscene or indecent words or language.
B. Transmissions to establish one-way communications.
C. Transmissions to establish model aircraft control.
D. Transmissions for third party communications.

Welcome to ham radio where you will hear most everything and anything over the airwaves. But one thing ***absolutely not allowed*** on any ham radio frequency, including repeaters, ***is indecent and obscene language***. And while a swear word here and there might get through, rank discussions with foul language are absolutely not tolerated on ham radio. [97.113(a)(4)] **ANSWER A.**

T2A11 What are the FCC rules regarding power levels used in the amateur bands?

A. Always use the maximum power allowed to ensure that you complete the contact.
B. An amateur may use no more than 200 watts PEP to make an amateur contact.
C. An amateur may use up to 1500 watts PEP on any amateur frequency.
D. An amateur must use the minimum transmitter power necessary to carry out the desired communication.

Be careful of this one – the answer is not a numerical one, but rather a philosophical one. ***Always run the minimum amount of power*** to make contact with another station. [97.313a]
ANSWER D.

Use the minimum amount of power output to make contact with another station.

T1D10 What is the meaning of the term broadcasting in the FCC rules for the amateur services?

A. Two-way transmissions by amateur stations.
B. Transmission of music.
C. Transmission of messages directed only to amateur operators.
D. Transmissions intended for reception by the general public.

You may not operate your station like an AM, FM or shortwave ***broadcast station***. You cannot transmit ***directly to the public***. [97.3(a)(10)] **ANSWER D.**

T1D09 Under which of the following circumstances are amateur stations authorized to transmit signals related to broadcasting, program production, or news gathering, assuming no other means is available?

A. Only where such communications directly relate to the immediate safety of human life or protection of property.
B. Only when broadcasting communications to or from the space shuttle.
C. Only where noncommercial programming is gathered and supplied exclusively to the National Public Radio network.
D. Only when using amateur repeaters linked to the Internet.

Ham radio signals get through when other radio services and cell phones may jam up. When the big hurricane formed across the southeast, ham radio stations and signals stayed up on the air. Normally, we are not permitted to assist in news gathering for our commercial television stations. However, in an emergency, where regular commercial radio broadcast station transmitters have been blown over, ham operators are permitted to assist in emergency news gathering, assuming there is absolutely no other radio system available, and assuming that their reports will assist the general public in ***protecting their lives and property***. [97.113(b)] **ANSWER A.**

T1D11 Which of the following types of communications are permitted in the Amateur Radio Service?

A. Brief transmissions to make station adjustments.
B. Retransmission of entertainment programming from a commercial radio or TV station.
C. Retransmission of entertainment material from a public radio or TV station.
D. Communications on a regular basis that could reasonably be furnished alternatively through other radio services.

Let's first look at what hams are NOT permitted to do – we are not permitted to hold a microphone up to an AM radio or TV speaker to re-transmit the signal on ham bands. We also are not allowed to transmit regular broadcasting that could take place on other radio services, such as marine ship-to-ship calls, land mobile base-to-mobile van calls, or pizza base to their local delivery units. One thing we MAY transmit are ***brief transmissions to make station adjustments***. Always be sure to announce your call letters, too. [97.113(a)(5)] **ANSWER A.**

Ham Hint: *Ham radio operators enjoy tinkering with their equipment. If you're at your workbench testing the quality of your transmissions by sending a signal into a non-radiating dummy load, there is still plenty of signal for you to monitor with another receiver. Try to avoid any continuous testing when hooked up to an outside antenna, unless you are testing the antenna itself.*

T1A04 Which of the following meets the FCC definition of harmful interference?

A. Radio transmissions that annoy users of a repeater.
B. Unwanted radio transmissions that cause costly harm to radio station apparatus.
C. That which seriously degrades, obstructs, or repeatedly interrupts a radio communications service operating in accordance with the Radio Regulations.
D. Static from lightning storms.

Amateur operators regularly conduct organized round-table gatherings on a specific frequency. This is called a "net," and "nets" are a fun way to meet new friends and get started on ham radio. If another amateur ***operator*** repeatedly ***transmits on a frequency*** that is already ***in use*** by members of the net, and if that operator does not stop transmitting when requested to do so by the net control, this would be considered ***harmful interference*** and is absolutely illegal. Luckily, occurrences like this bad behavior are relatively rare on the ham bands. And the very best procedure to discourage harmful interference is to simply ignore the interfering station and make absolutely no reference that someone is out there trying to break up the net. Without recognition, they probably will go away. [97.3(a)(23)] **ANSWER C.**

T1D03 When is the transmission of codes or ciphers allowed to hide the meaning of a message transmitted by an amateur station?

A. Only during contests.
B. Only when operating mobile.
C. Only when transmitting control commands to space stations or radio control craft.
D. Only when frequencies above 1280 MHz are used.

This question asks about secret codes and ciphers – intentional scrambling to prevent eavesdropping. This normally is never allowed, except for those control operators who are ***transmitting special telecommands to ham radio satellites***. Also, hams who enjoy ***flying their model aircraft*** using digital code encryption are permitted to do so to ensure their aircraft doesn't accidentally get the wrong commands from another transmitter. But here is something new – we are beginning to see some "leaks" in proper ham radio distribution, with equipment coming into the US directly from overseas, intended for the land-mobile market, but which also works quite nicely on ham frequencies, too. This cheap gear may include voice encryption, and this is not permitted in the ham radio service, so steer clear of imported, no-name ham gear, and never use any type of voice encryption on the ham bands. Check to make sure the equipment you are buying is FCC Certified before handing over your hard-earned cash. [97.113(a)(4), 97.211(b), 97.217] **ANSWER C.**

T1D04 What is the only time an amateur station is authorized to transmit music?

A. When incidental to an authorized retransmission of manned spacecraft communications.
B. When the music produces no spurious emissions.
C. When the purpose is to interfere with an illegal transmission.
D. When the music is transmitted above 1280 MHz.

Music is generally not allowed on the ham bands. No playing the violin or piano, and no singing happy birthday. However, a little known rule could permit you to blow your trumpet for reveille when sending up an authorized signal to the space shuttle or international space station. This music is considered "***incidental to an authorized retransmission of manned spacecraft communications***." You do play the trumpet, right? [97.113(a)(4), 97.113(e)] **ANSWER A.**

T1D05 When may amateur radio operators use their stations to notify other amateurs of the availability of equipment for sale or trade?

A. When the equipment is normally used in an amateur station and such activity is not conducted on a regular basis.
B. When the asking price is $100.00 or less.
C. When the asking price is less than its appraised value.
D. When the equipment is not the personal property of either the station licensee or the control operator or their close relatives.

Once a week you tune into an interesting, fun radio net where hams, in turn, give their call sign and list a piece of ham radio equipment they own that's up for sale. The ***occasional sale of your OWN ham radio equipment is not considered a business rule violation***, as long as you don't buy and sell ham radio equipment regularly for profit. So these "swap nets" are perfectly okay, as long as your offering of equipment is just on an occasional basis. [97.113(a)3] **ANSWER A.**

T1F13 When must the station licensee make the station and its records available for FCC inspection?

A. Any time upon request by an official observer.
B. Any time upon request by an FCC representative.
C. 30 days prior to renewal of the station license.
D. 10 days before the first transmission.

While it rarely happens to ham operators sticking to the law, an ***FCC knock on the front door requires you to let them in for an inspection*** of your station equipment and station records. It usually doesn't happen out of the blue – if you are on the air and regularly wiping out TV reception for the entire block, the FCC may make contact with you and ask you to investigate what is going on. Who knows – it might be a leaky cable problem out your back door. But if the problem persists, the FCC could ask to inspect your station. They will not take NO for an answer! [97.103(c)] **ANSWER B.**

T1C07 What may result when correspondence from the FCC is returned as undeliverable because the grantee failed to provide the correct mailing address?

A. Fine or imprisonment.
B. Revocation of the station license or suspension of the operator license.
C. Require the licensee to be re-examined.
D. A reduction of one rank in operator class.

The Federal Communications Commission licensing bureau needs an up-to-date postal mailing address. Where do you want the official license to be sent? A P.O. Box is fine, or maybe at Mom's house, but it must be a postal address that would never refuse correspondence from the FCC. If the FCC ever gets back any correspondence as undeliverable, it could quickly ***suspend your ham radio license***. If you decide to move your mailing address, you may log on to the Universal Licensing System and update your mailing address. Working with the ULS can be a bit complicated – it asks for your Federal Registration Number (FRN) and password that will be assigned to you when your license is granted. Be sure to save all FCC correspondence when your license arrives because it may contain your password to give you access to ULS. An easier way to keep your mailing address up-to-date is to contact The W5YI VEC at 800-669-9594, and take advantage of their change-of-address service. [97.23] **ANSWER B.**

▾ IF YOU'RE LOOKING FOR	▾ THEN VISIT
USA ITU Info	www.itu.int/home/
Full Part 97	www.ARRL.org/FandES/Field/Regulations/io
Entire Question Pools	www.ncvec.org/
Purchase a Printed Copy of Part 97	www.w5yi.org

T5C06 What is the abbreviation that refers to radio frequency signals of all types?

A. AF. C. RF.
B. HF. D. VHF.

If a unit transmits, it puts out a radio frequency. A wireless hot spot has a radio frequency. Your cell phone has a radio frequency output. Your new Technician Class transceiver has a radio frequency output, as soon as you push the transmit button. The ***term "RF" refers to radio frequency***. **ANSWER C.**

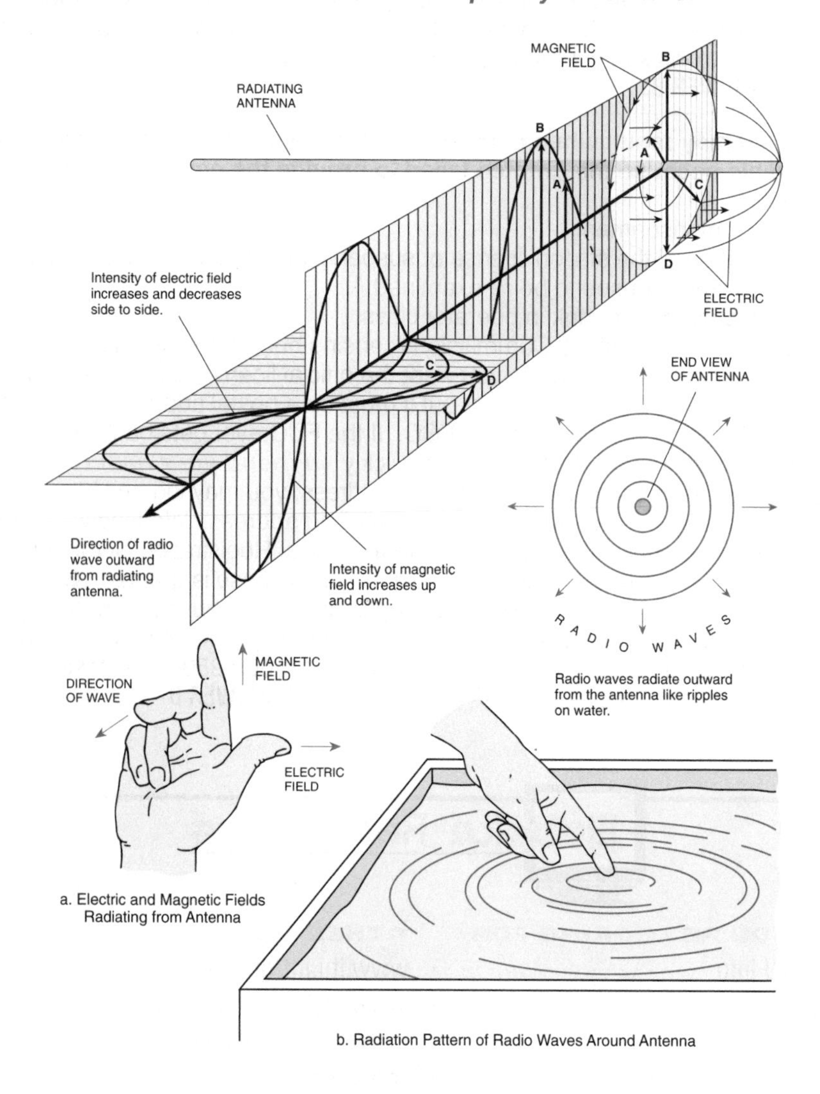

a. Electric and Magnetic Fields Radiating from Antenna

b. Radiation Pattern of Radio Waves Around Antenna

Radio Waves

Source: Basic Electronics © 1994, 2000 Master Publishing, Inc., Niles, Illinois

T3A07 What type of wave carries radio signals between transmitting and receiving stations?

A. Electromagnetic.
B. Electrostatic.
C. Surface acoustic.
D. Magnetostrictive.

Radio waves contain both electrical and magnetic energy in fields that are at right angles to each other. The combination of both electrical and magnetic energy is called ***electromagnetic*** energy. That is a ***radio wave***! **ANSWER A.**

T3B03 What are the two components of a radio wave?

A. AC and DC.
B. Voltage and current.
C. Electric and magnetic fields.
D. Ionizing and non-ionizing radiation.

The ***electric and magnetic fields*** of a radio wave are at right angles to each other, and together they are called "electromagnetic" radio waves. **ANSWER C.**

T3B04 How fast does a radio wave travel through free space?

A. At the speed of light.
B. At the speed of sound.
C. Its speed is inversely proportional to its wavelength.
D. Its speed increases as the frequency increases.

The velocity of ***radio waves*** is the same as the ***speed of light***: 300,000,000 meters per second. 300 million meters per second is the same velocity as the speed of light in a vacuum. Oh yeah, radio waves slow down a bit as they go through the ionosphere, clouds, and our smog layers, but except for G.P.S. calculations, you wouldn't know it! **ANSWER A.**

T3B11 What is the approximate velocity of a radio wave as it travels through free space?

A. 3000 kilometers per second.
B. 300,000,000 meters per second.
C. 300,000 miles per hour.
D. 186,000 miles per hour.

The velocity of radio waves through space is ***300,000,000 meters per second***, referred to as 300 hundred million meters per second. **ANSWER B.**

T5C05 What is the unit of frequency?

A. Hertz.
B. Henry.
C. Farad.
D. Tesla.

The basic unit of frequency is the ***hertz***, abbreviated Hz. **ANSWER A.**

Ham Hint: *Radio waves are invisible, yet at a high enough microwave frequency, you could actually feel them! So keep your hands out of the turned-on microwave oven! We generally consider anything above 20,000 cycles per second (Hertz) as radio frequencies, although the Navy has some radio transmitters that go lower than this. We sometimes abbreviate 20,000 as 20 kilohertz (kHz).*

T5B07 If a frequency readout calibrated in megahertz shows a reading of 3.525 MHz, what would it show if it were calibrated in kilohertz?

A. 0.003525 kHz.
B. 35.25 kHz.
C. 3,525 kHz.
D. 3,525,000 kHz.

Move the decimal point 3 places to the right to convert MHz to kHz. 3.525 MHz is ***3,525 kHz***. Most HF radios display frequency in kilohertz. **ANSWER C.**

T3B01 What is the name for the distance a radio wave travels during one complete cycle?

A. Wave speed.
B. Waveform.
C. Wavelength.
D. Wave spread.

The key word here is distance that a radio wave travels over one complete cycle. We measure the distance radio signals travel as wavelength. Keywords are ***DISTANCE and WAVELENGTH***. **ANSWER C.**

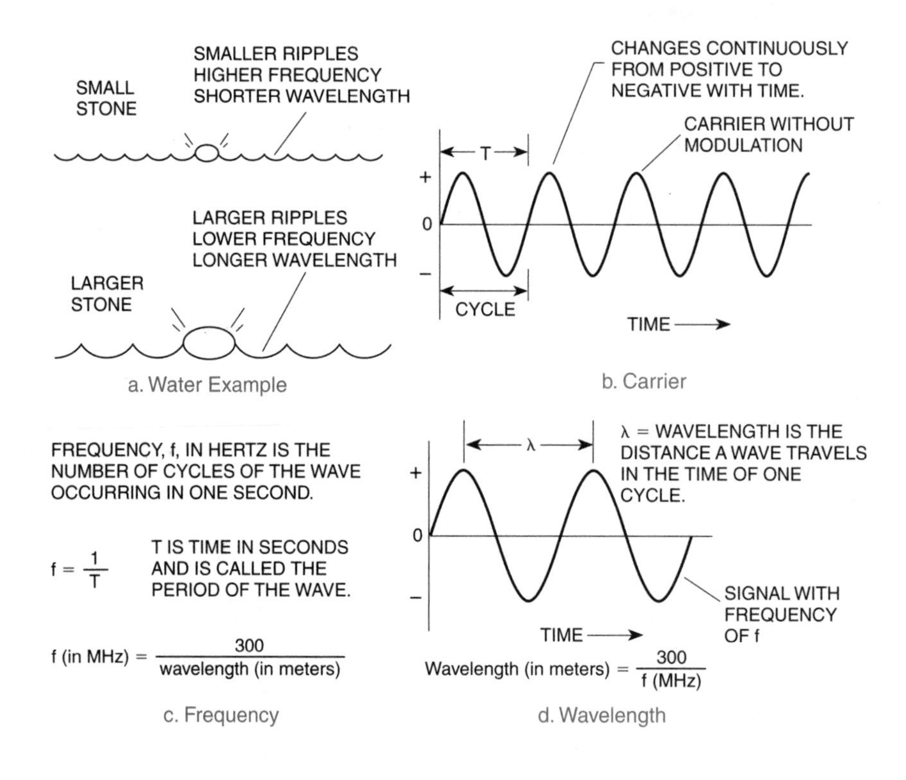

Carrier, Frequency, Cycle and Wavelength

T3B07 What property of radio waves is often used to identify the different frequency bands?

A. The approximate wavelength.
B. The magnetic intensity of waves.
C. The time it takes for waves to travel one mile.
D. The voltage standing wave ratio of waves.

When we want to meet a pal on a specific ham band, we usually say which band it is in meters. The most popular ham band for Technician Class operators is the 2-meter band. Your pal says that she has a 2-meter radio, so the next question is, what exact FREQUENCY? So it's really a toss up on whether or not you want to tell someone to meet you at a specific frequency, or ask them whether or not they have a 2-meter radio, which describes the ***WAVELENGTH of the band*** on which you plan to operate. So if you asked me, I would first determine what wavelength band we were going to operate on, and then indicate the specific frequency we should try to make contact on. **ANSWER A.**

T3B05 How does the wavelength of a radio wave relate to its frequency?

A. The wavelength gets longer as the frequency increases.
B. The wavelength gets shorter as the frequency increases.
C. There is no relationship between wavelength and frequency.
D. The wavelength depends on the bandwidth of the signal.

The ***higher*** we go in ***frequency***, the ***shorter*** the ***distance*** between each wave. The LOWER we go in frequency, the LONGER the distance between each wave. Now say this out loud – LOWER LONGER, HIGHER SHORTER. Got it? Now look around and see who is staring at you wondering what in the world you are talking about! **ANSWER B.**

T3B06 What is the formula for converting frequency to wavelength in meters?

A. Wavelength in meters equals frequency in hertz multiplied by 300.
B. Wavelength in meters equals frequency in hertz divided by 300.
C. Wavelength in meters equals frequency in megahertz divided by 300.
D. Wavelength in meters equals 300 divided by frequency in megahertz.

Frequency in MHz and wavelength in meters are inversely proportional. The root number to remember is 300! Wavelength is the distance a wave travels in the time of one cycle. Wavelength usually is stated in meters or centimeters.
Frequency in Hertz is the number of cycles of the wave occurring in one second.

F(in MHz) = 300 ÷ wavelength (in meters)
Wavelength (in meters) = 300 ÷ F (MHz)

On your upcoming Technician class examination, the above 4 answers will be identical on the test, and they read somewhat similar. Just remember: 300 ÷ frequency in MHz. This correct answer is the ONLY answer that ends with the word "megahertz." **ANSWER D.**

Conversions Between Wavelength and Frequency

Converting Frequency to Wavelength

To find wavelength (λ) in meters, if you know frequency (f) in megahertz (MHz), Solve:

$$\lambda(\text{meters}) = \frac{300}{f(\text{MHz})}$$

Converting Wavelength to Frequency

To find frequency (f) in megahertz (MHz), if you know wavelength (λ) in meters, Solve:

$$f(\text{MHz}) = \frac{300}{\lambda(\text{meters})}$$

T3B10 What frequency range is referred to as HF?

A. 300 to 3000 MHz.
B. 30 to 300 MHz.
C. 3 to 30 MHz.
D. 300 to 3000 kHz.

High frequency, called ***HF, extends from 3 MHz to 30 MHz.*** As a new Technician Class operator, you have 4 bands of high frequency operation. HF offers exciting daytime and nighttime skywave "skip" conditions. There is no longer a Morse code test required to operate on your 4 Technician Class high frequency sub-bands. You will need to know Morse code to take advantage of your Morse-code-only privileges on HF 80-, 40-, 15- and 10-meter bands. But good news – on 10 meters, you have voice single sideband privileges from 28.3 to 28.5 MHz, and during the summertime we get plenty of daytime and late afternoon skip, letting you yak with fellow Technician Class operators all over the country! And, since we are on the solar cycle increase, Solar Cycle 24 may even give you some voice contacts on 10 meters to Europe, South America, and Asia! As a brand new Technician Class operator, you may own and operate a large, high-frequency ham radio station, including a major-sized directional antenna system, too! **ANSWER C.**

T3B08 What are the frequency limits of the VHF spectrum?

A. 30 to 300 kHz.
B. 30 to 300 MHz.
C. 300 to 3000 kHz.
D. 300 to 3000 MHz.

The ***VHF*** spectrum extends from ***30 MHz to 300 MHz***. **ANSWER B.**

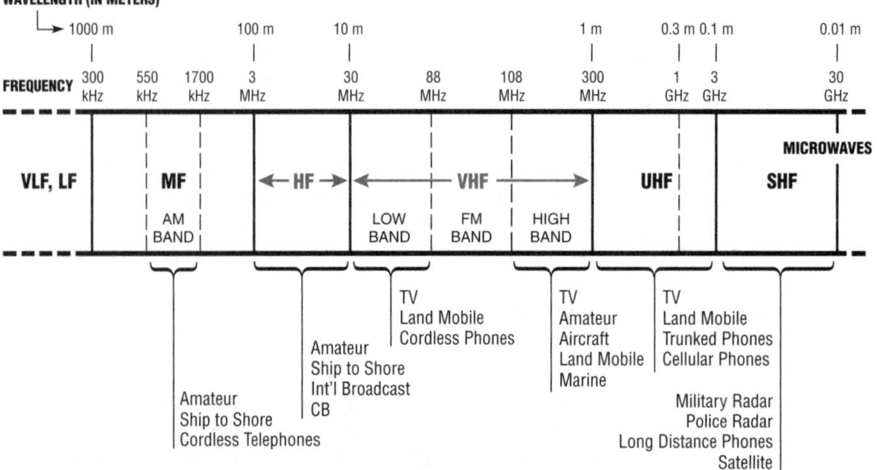

As Frequency increases, wavelength becomes shorter, as you can see from this radio frequency spectrum chart. Transmissions that require greater bandwidth, such as TV, use higher frequencies.

T1B03 Which frequency is within the 6 meter band?

A. 49.00 MHz.
B. 52.525 MHz.
C. 28.50 MHz.
D. 222.15 MHz.

52.525 MHz is right in the middle of our 6-meter band, which extends from 50 to 54 MHz. [97.301(a)] **ANSWER B.**

CW

50 MHz 50.1 MHz 54 MHz

6-Meter Wavelength Band Privileges

T1B04 Which amateur band are you using when your station is transmitting on 146.52 MHz?

A. 2 meter band.
B. 20 meter band.
C. 14 meter band.
D. 6 meter band.

146.52 MHz is right in the middle of our ***2-meter band***, which extends from 144 to 148 MHz. [97.301(a)] **ANSWER A.**

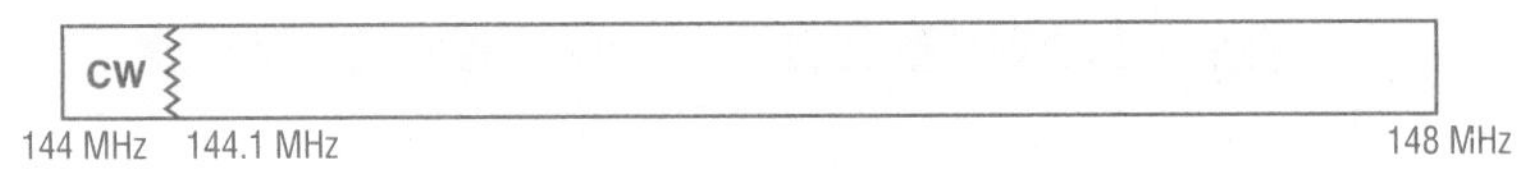

2 Meter Wavelength Band Privileges

T1B10 Which of the bands available to Technician Class operators have mode-restricted sub-bands?

A. The 6 meter, 2 meter, and 70 cm bands.
B. The 2 meter and 13 cm bands.
C. The 6 meter, 2 meter, and 1.25 meter bands.
D. The 2 meter and 70 cm bands.

The ***three bands*** that ***have sub-band restrictions*** to designated modes of operation are the ***6-meter, 2-meter, and the 1 1/4-meter bands***. The term "mode restricted sub-band" means no handheld or mobile FM radio transmitting here!

6 meters	50.0 – 50.1	No FM!
2 meters	144.0 – 144.10	No FM!
1.25 meters	222.0 – 222.34	No FM!

Go with answer C for the examination, but understand that each and every ham band has certain areas where handheld, base, and mobile FM (frequency modulation) is discouraged. This is why I always recommend getting your new multi-band mobile or handheld transceiver pre-programmed by the dealer or your local club technical officer who knows how to "clone" a full set of FM frequencies, leaving mode-restricted frequencies out of your FM lineup. [97.305(c)] **ANSWER C.**

T1B11 What emission modes are permitted in the mode-restricted sub-bands at 50.0 to 50.1 MHz and 144.0 to 144.1 MHz?

A. CW only.
B. CW and RTTY.
C. SSB only.
D. CW and SSB.

The term "sub-band" refers to an area within a ham band that is restricted to very narrow bandwidth emission types, specifically Morse code. Morse code is abbreviated CW, for continuous wave. 50.0-50.1 MHz on the bottom of the 6-meter band, and 144.0-144.1 MHz on the bottom of the 2-meter band are areas reserved for CW only. RTTY is radio teleprinter – too wide – and same thing for SSB, single sideband, too wide. ***CW only on these two sub-bands***. [97.305 (a)(c)] **ANSWER A.**

Ham Hint: *Many weak-signal operators use CW on the bottom of the 2-meter sub-band to bounce signals off of the moon, and off of auroras! Hear for yourself all of the excitement of moonbounce, aurora, and long-haul tropospheric ducting on the audio CD included in the front of this book. There's a lot of ham radio fun to be had just with your Technician license!*

T1B07 What amateur band are you using if you are transmitting on 223.50 MHz?

A. 15 meter band.
B. 10 meter band.
C. 2 meter band.
D. 1.25 meter band.

The ***223 MHz*** band is more commonly called 1.25 meters and sometimes 1-1/4 meters. Check this out by using the equation in the Elmer Hint on page 57. The calculator keystrokes are: "Clear 300 ÷ 223.50 = ." [97.301(a)] **ANSWER D.**

222 MHz 225 MHz

1.25 Meter Wavelength Band Privileges

T8D05 Which of the following emission modes may be used by a Technician Class operator between 219 and 220 MHz?

A. Spread spectrum.
B. Data.
C. SSB voice.
D. Fast-scan television.

Use it or lose it! In years past, our 1 1/4-meter band was 5-MHz wide, extending from 220 to 225 MHz. Due to perceived inactivity by the FCC on our use of the lower portion of the band, WE LOST 220-222 MHz. But we did recover 1 MHz of prime "radio real estate" at ***219 to 220 MHz for point-to-point digital message forwarding***. This is the "backbone" of many digital wireless links in most areas of the country, except for the Mississippi River region, where river traffic has priority use of this band. **ANSWER B.**

T3B09 What are the frequency limits of the UHF spectrum?

A. 30 to 300 kHz.
B. 30 to 300 MHz.
C. 300 to 3000 kHz.
D. 300 to 3000 MHz.

The ***UHF*** spectrum extends from ***300 MHz to 3000 MHz***. Watch out for answer C – this incorrectly lists kHz, not MHz! **ANSWER D.**

T1B05 Which 70 cm frequency is authorized to a Technician Class license holder operating in ITU Region 2?

A. 53.350 MHz.
B. 146.520 MHz.
C. 443.350 MHz.
D. 222.520 MHz.

443.350 MHz is right in the middle of the 70 cm band, where our privileges extend from 420 up to 450 MHz. [97.301(a)] **ANSWER C.**

T2A02 What is the national calling frequency for FM simplex operations in the 70 cm band?

A. 146.520 MHz.
B. 145.000 MHz.
C. 432.100 MHz.
D. 446.000 MHz.

The national calling ***FM simplex frequency on the 70 cm band***, referred to as the 440 MHz band, is ***446.000 MHz***. Watch out for answer A, that's the national simplex for the 2 meter band. Go with answer D. **ANSWER D.**

420 MHz 450 MHz

70-CM Meter Wavelength Band Privileges

T1B06 Which 23 cm frequency is authorized to a Technician Class operator license?

A. 2315 MHz.
B. 1296 MHz.
C. 3390 MHz.
D. 146.52 MHz.

The ***23-cm band extends from 1240 to 1300 MHz***, and is a good one for amateur television, repeater operation, and point-to-point simplex where there is plenty of ready-made equipment for this band. 1296 MHz is near the top end of our 23-CM band privileges. If you buy a "tri-band" transceiver, chances are it will have 2 meters, 440 MHz, and full capabilities for the 23-cm band from 1240 to 1300 MHz. [97.301(a)] **ANSWER B.**

1240 MHz | 1300 MHz

23-CM Wavelength Band Privileges

T2A10 What is a band plan, beyond the privileges established by the FCC?

A. A voluntary guideline for using different modes or activities within an amateur band.
B. A mandated list of operating schedules.
C. A list of scheduled net frequencies.
D. A plan devised by a club to use a frequency band during a contest.

Every amateur radio band is sliced up into specific operating band plans. ***Many of the band plans are voluntary guidelines*** that hams have established for specific data signals, voice operation, automatic position reporting system operation, weak signal work, DXing, slow-scan amateur television, propagation beacons, and specific areas for working satellites and the International Space Station and the Space Shuttle. **ANSWER A.**

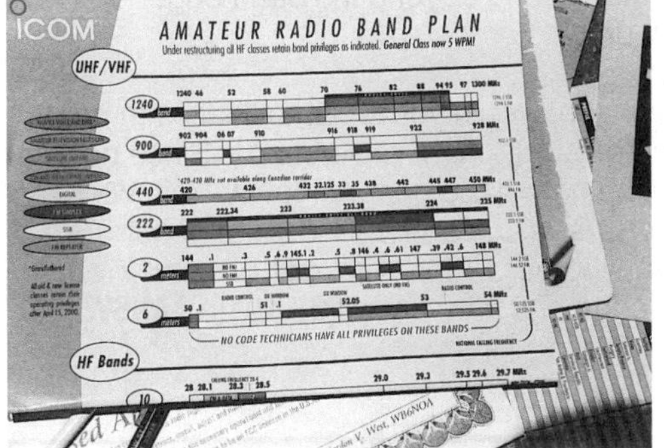

Band plans subdivide ham radio bands for specific uses, such as data, repeaters, and weak signal CW.

For a free band plan chart visit
➡ **www.icomamerica.com**

Ham Hint: *It takes time to understand where the 2-meter and 70 cm band plans are, so when you get your new equipment, let a fellow ham help you program it to insure you operate within the voluntary band plan. In other words, don't just buy a brand new handheld and start talking with your class buddy on the frequency that just happens to come up on the display when you first turn it on. Get some local help to make sure you begin operating within the voluntary band plan.*

T1B08 What do the FCC rules mean when an amateur frequency band is said to be available on a secondary basis?

A. Secondary users of a frequency have equal rights to operate.
B. Amateurs are only allowed to use the frequency at night.
C. Amateurs may not cause harmful interference to primary users.
D. Secondary users are not allowed on amateur bands.

We share the 900-MHz band with the vehicle locator service, which is the ***primary user*** of the frequencies. Same thing with 70 cm – we share it with military radio location services. They ***have first rights to these frequencies***. [97.303] **ANSWER C.**

T1C05 What must you do if you are operating on the 23 cm band and learn that you are interfering with a radiolocation station outside the United States?

A. Stop operating or take steps to eliminate the harmful interference.
B. Nothing, because this band is allocated exclusively to the amateur service.
C. Establish contact with the radiolocation station and ask them to change frequency.
D. Change to CW mode, because this would not likely cause interference.

While this may be a rare occurrence, FM phone communications in the 23-cm band could cause interference to a radiolocation station outside of the United States, and if you are informed you are ***causing interference***, you should ***stop operating*** immediately. [97.303(h)] **ANSWER A.**

T1B09 Why should you not set your transmit frequency to be exactly at the edge of an amateur band or sub-band?

A. To allow for calibration error in the transmitter frequency display.
B. So that modulation sidebands do not extend beyond the band edge.
C. To allow for transmitter frequency drift.
D. All of these choices are correct.

You wouldn't sit on a roof ledge 100 stories up, would you? Same thing with ham radio operation – we don't operate on the edge of a ham radio band. Since none of our radio emissions are allowed beyond the ham band upper or lower limit, we stay well within band edges, ***in case our signals should drift***, or some of our ***signal extends beyond the band edge***, or maybe our ***radio was not properly calibrated***. Never operate right on the edge of a ham band! [97.101(a)] **ANSWER D.**

▾ IF YOU'RE LOOKING FOR	▾ THEN VISIT
Ham Band Plans and Operating Frequencies	www.ac6v.com/frequencies.htm
U.S. Amateur Radio Frequency Allocations	www.ARRL.org/FandES/Field/Regulations/allocate.html

CD 2 TRACK 1

Your First Radio

T4B04 What is a way to enable quick access to a favorite frequency on your transceiver?

A. Enable the CTCSS tones.
B. Store the frequency in a memory channel.
C. Disable the CTCSS tones.
D. Use the scan mode to select the desired frequency.

There are literally hundreds of hot frequencies in most major cities in the US. As a new ham, looking at a frequency guide alone won't necessarily give you a clue as to which ones are great stations to tune in and join in on conversations. I recommend that you purchase your handheld through a local authorized ham radio dealer who might store about 10 of the hot local channels in the memory of your radio. Mail order companies may do this, too, as long as they know where the action is on the radio dial in your area. ***Storing popular frequencies in a memory channel*** will allow you to quickly go from one repeater to another, ready to transmit if the conversation sounds inviting! **ANSWER B.**

With a transceiver like one of these, you can hold your ham station in the palm of your hand.

T9A04 What is a disadvantage of the "rubber duck" antenna supplied with most handheld radio transceivers?

A. It does not transmit or receive as effectively as a full-sized antenna.
B. It transmits a circularly polarized signal.
C. If the rubber end cap is lost it will unravel very quickly.
D. All of these choices are correct.

When you open up the box containing your dual-band handheld, you will find a small "wall wart" battery charger, the rechargeable battery pack or a battery tray, an instruction manual, and the famous "rubber duck" antenna. This antenna is the bare minimum in getting a good signal out on the airwaves. It's okay for transmitting simplex, direct from you to a buddy a mile away, but for reaching out to a repeater more than 5 miles away, the stock, factory-supplied ***rubber duck antenna*** is a compromise and ***is not as effective as a full sized antenna***. When you buy your new dual-band handheld, consider one of those flexible, 15-inch, dual-band antennas. It will do well out in the open. And for vehicle use, consider a magnetic antenna as well. That magnetic-mount vehicle antenna comes with coax cable, and you need to be sure to get the right connector on the end of the cable that will screw on or twist on to your new handheld. The vehicle antenna is full-sized, but the rubber duck is not! **ANSWER A.**

T9A07 What is a good reason not to use a "rubber duck" antenna inside your car?

A. Signals can be significantly weaker than when it is outside of the vehicle.
B. It might cause your radio to overheat.
C. The SWR might decrease, decreasing the signal strength.
D. All of these choices are correct.

If you attempt to transmit with the rubber duck antenna inside your vehicle, your signal will barely make it out the windows. All that metal will make your ***signal 10 to 20 times weaker***. For inside the vehicle, you MUST use an external magnetic-mount antenna with that brand new dual-band radio. Then, later on, you can get a hatch-lip-mount antenna for your car. **ANSWER A.**

Modern dual- and tri-band handheld transceivers like these have amazing built-in capabilities that make ham radio easy, fun, and portable!

T7A10 What device increases the low-power output from a handheld transceiver?

A. A voltage divider.
B. An RF power amplifier.
C. An impedance network.
D. A voltage regulator.

As a new Technician Class operator, you will probably choose a handheld, dual-band as your first radio. You can run this in your vehicle using an outside antenna with a magnetic mount and achieve great results! I would also suggest a filtered DC adaptor plug and a headset speaker/mic, too. If you really need more than the 5 watts of power that comes out of the handheld, you also could purchase a ***power amplifier*** that would boost the 5 watts to up to 100 watts power output. But I think you'll find the outside antenna may be all you'll need for successful handheld operation from inside your vehicle. **ANSWER B.**

T8A04 Which type of modulation is most commonly used for VHF and UHF voice repeaters?

A. AM.
B. SSB.
C. PSK.
D. FM.

We use ***frequency modulation (FM)*** on most VHF and UHF repeaters. **ANSWER D.**

Ham Hint: *Ask the radio seller to please preprogram at least 10 memory channels before you take delivery of your brand-new, dual-band radio. This way, you'll have some popular repeaters ready to go as soon as you turn it on!*

T8A09 What is the approximate bandwidth of a VHF repeater FM phone signal?

A. Less than 500 Hz .
B. About 150 kHz.
C. Between 5 and 15 kHz.
D. Between 50 and 125 kHz.

When we operate our 2-meter, 1-1/4-meter, and 70 cm handheld using frequency modulation, ***the FM signal is between 5 and 15 kHz wide***. With properly adjusted equipment, 10 kHz total FM bandwidth is about normal. **ANSWER C.**

T8A02 What type of modulation is most commonly used for VHF packet radio transmissions?

A. FM.
B. SSB.
C. AM.
D. Spread Spectrum.

The majority of VHF and UHF communications incorporates frequency modulation, FM. This is the same mode that we use for ***VHF data packet*** communications, too, *FM*. **ANSWER A.**

T4A01 Which of the following is true concerning the microphone connectors on amateur transceivers?

A. All transceivers use the same microphone connector type.
B. Some connectors include push-to-talk and voltages for powering the microphone.
C. All transceivers using the same connector type are wired identically.
D. Un-keyed connectors allow any microphone to be connected.

When we speak into a microphone, sound pressure waves strike an element which converts acoustic energy into electrical waves. The electric waves then lead to the transmitter that will ultimately transmit electromagnetic waves onto the airwaves. Microphone jacks and power connections on handhelds as well as mobile equipment may look similar between different manufacturers, but they are not! ***Some mobile radio microphone jacks output a small DC voltage*** to power an illuminated keypad. Each manufacturer assigns this voltage to a specific pin, and if you accidentally plug in the wrong microphone, you could short the voltage to ground, zapping your radio instantly. Its about a $125 fix! Always make sure the microphone, or custom Heil headset, has the correct adapter cable for your specific radio brand and model number. There are many advantages to going with professional headsets, but make sure you order the correct adapter cable!
ANSWER B.

T4A02 What could be used in place of a regular speaker to help you copy signals in a noisy area?

A. A video display.
B. A low pass filter.
C. A set of headphones.
D. A boom microphone.

When performing public service work in a ***noisy environment, headphones*** or a simple earphone is a great way to hear audio coming out of your radio's receiver.
ANSWER C.

T4A03 Which is a good reason to use a regulated power supply for communications equipment?

A. It prevents voltage fluctuations from reaching sensitive circuits.
B. A regulated power supply has FCC approval.
C. A fuse or circuit breaker regulates the power.
D. Power consumption is independent of load.

Never try to run your 12-volt equipment directly from a battery charger, or attempt to transmit with your handheld when it is plugged into the wall adaptor. If you plan to run your equipment off of commercial power with it still plugged in, always use a ***regulated power supply to protect your gear from voltage fluctuations***. The greatest cause of handheld transmit hum, and sometimes damage to the equipment, is trying to transmit with it still plugged into the AC receptacle power adaptor. **ANSWER A.**

A regulated power supply like this one provides constant voltage to your radio, protecting it from voltage fluctuatios.

T8A03 Which type of voice modulation is most often used for long-distance or weak signal contacts on the VHF and UHF bands?

A. FM. C. SSB.
B. AM. D. PM.

Unlike the common FM mode that you will be using with your dual-band handheld, ***single sideband (SSB)*** requires special – yet not too expensive – equipment, and 6-meter single-sideband radios are seen regularly at swap meets. Six-meter SSB signals are ideal for bouncing off the ionosphere. And, as you will hear on the enclosed audio CD, it is mighty exciting! **ANSWER C.**

T8A07 What is the primary advantage of single sideband over FM for voice transmissions?

A. SSB signals are easier to tune.
B. SSB signals are less susceptible to interference.
C. SSB signals have narrower bandwidth.
D. All of these choices are correct.

We use frequency modulation through FM repeaters on many popular VHF and UHF ham bands. But near the bottom of 6 meters, 2 meters, 1 1-4 meters, and at 432 and 1296 MHz, we use ***single sideband***, which ***uses less bandwidth than FM signals***. **ANSWER C.**

Ham Hint: *Handhelds do not incorporate a single-sideband capability, but some "multimode" mobiles do include SSB operation. Listen to my CD and discover the fascinating world of longer range transmissions over hundreds and sometimes thousands of miles using single sideband. Listen to the regular summertime occurrences of long-range skywave SSB signals on the 6- and 10-meter bands! After you listen to the included audio CD, hopefully you will get very excited about all the fun that single sideband can add to your ham radio activity beyond your daily communications on regular FM repeaters.*

T8A06 Which sideband is normally used for 10 meter HF, VHF and UHF single-sideband communications?

A. Upper sideband.
B. Lower sideband.
C. Suppressed sideband.
D. Inverted sideband.

On VHF and UHF, ***we always use upper sideband*** as our communications mode. Single sideband is an advanced feature after you have been on the air for a few months as a Technician Class operator. You won't find single sideband capability in a ham handheld. You will find SSB in all worldwide high frequency radio sets. This equipment may also give you 6 meters, 2 meters, and 70 cm, in addition to the worldwide HF capability. We don't use single sideband through repeaters. SSB is found at the bottom of most VHF/UHF ham bands, so you will want to try the following frequencies – all upper sideband – to hear some SSB signals:

10 meters	28.400 MHz SSB
6 meters	50.125 MHz SSB
2 meters	144.200 MHz SSB
70 cm	432.100 MHz SSB

No FM allowed on these weak-signal SSB frequencies. Worldwide equipment, which includes all of these bands plus HF, runs under $1,000. **ANSWER A.**

T7A06 What device takes the output of a low-powered 28 MHz SSB exciter and produces a 222 MHz output signal?

A. High-pass filter.
B. Low-pass filter.
C. Transverter.
D. Phase converter.

Many folks become ham radio operators from their experience on CB, 27 MHz. Some of their 27 MHz equipment may be modified and tied into a device called a ***"transverter,"*** making the 11 meter signal now coming out on 222 MHz, and signals coming in to your antenna on 222 MHz end up on 11 meters, thanks to the transverter. So if you have some older CB radio, multimode, don't toss it just yet. You might be able to use it on ham frequencies with a transverter. **ANSWER C.**

Gordon uses a 10 GHz transverter that down-converts the received signal to 144 MHz into his weak-signal, multi-mode radio for matitime microwave communications.

Website Resources

IF YOU'RE LOOKING FOR	THEN VISIT
Ham Equipment Reviews	www.EHAM.net
Ham Radio Outlet (HRO)	www.HamRadio.com
Amateur Electronics Supply (AES)	www.AESHAM.com
Largest Ham Accessory Catalog	www.MFJEnterprises.com
Advanced Specialties	www.advancedspecialties.net
Alltronics	www.alltronics.com
Amateur Accessories	www.amateuraccessories.com
Asscociated Radio	www.associatedradio.com
Austin Amateur Radio	www.aaradio.com
Cedar City Sales	www.cedarcitysales.com
Central Utah Electronics Supply	www.electronicspro.com
Communications Products	www.commproducts.net
GigaParts, Inc.	www.gigaparts.com
Houston Amateur Radio Supply	www.texasparadise.com/hars
Iowa Radio Supply Co., Inc	www.irsupply.com
Jubilee Enterprises	www.shopjubilee.com
Jun's Electronics	www.hamcity.com
K1CRA Radio Webstore	www.k1cra.com
K-Comm, Inc. – The Ham Store	www.kcomm.biz
KJI Electronics, Inc.	www.kjielectronics.com
Lentini Communications, Inc.	www.lentinicomm.com
Metal & Cable	www.metal-cable.com
R & L Electronics	www.randl.com
Rad-Comm Radio	www.radcomm.bizland.com/rad-comm
Radio City	www.radioinc.com
Rayfield Communications	www.rayfield.net
The Ham Station	www.hamstation.com
Universal Radio Inc	www.universal-radio.com
Waypoint Crusing Solutions	www.waypoints.com
WBOW, Inc.	www.wbow.com

Note: This list includes many ham radio dealers where you can purchase your first radio and accessories

Going On The Air

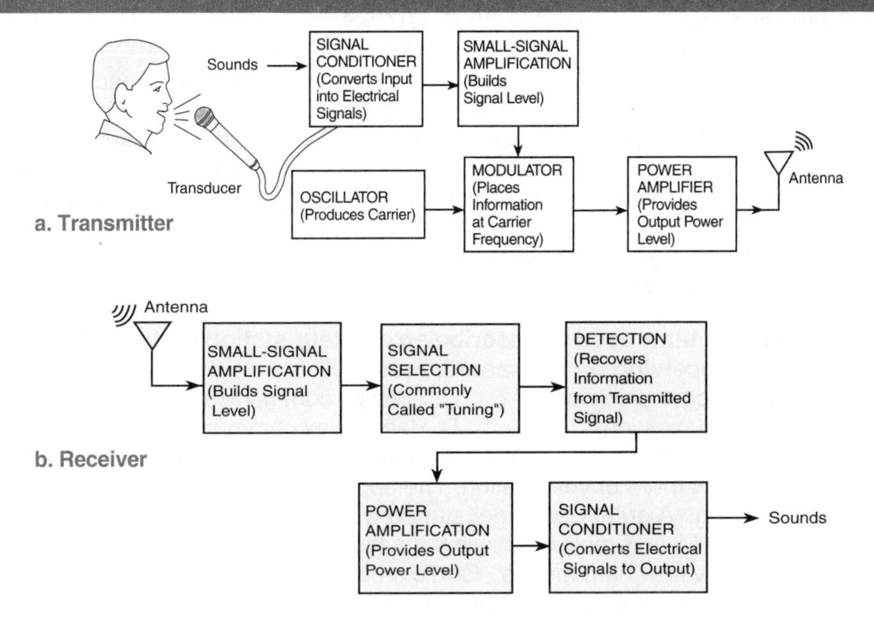

Block Diagram of a Basic Radio Communications System
Source: *Basic Communications Electronics*, Hudson & Luecke,
© 1999, Master Publishing, Inc., Niles, Illinois

T4B02 Which of the following can be used to enter the operating frequency on a modern transceiver?

A. The keypad or VFO knob.
B. The CTCSS or DTMF encoder.
C. The Automatic Frequency Control.
D. All of these choices are correct.

Most new Technician Class hams start with a dual-band handheld – the popular bands are 2 meters and 70 centimeters. The radio will have a ***keypad*** as well as a ***variable frequency oscillator*** (VFO) knob or up-and-down push buttons. The knob, push buttons, and keypad are a great way ***to select the frequency*** of choice. **ANSWER A.**

T4B03 What is the purpose of the squelch control on a transceiver?

A. To set the highest level of volume desired.
B. To set the transmitter power level.
C. To adjust the automatic gain control.
D. To mute receiver output noise when no signal is being received.

The ***squelch control*** on your FM radio equipment ***silences the background noise*** if there is no station on the air. As soon as your equipment detects a signal on the air, the squelch circuit will immediately open and the station you are receiving will magically be heard out of the speaker. Most dual-band handheld squelch circuits may be adjusted by a squelch knob. First rotate the volume to mid-range, and then turn the squelch control knob counterclockwise so you hear background noise when no one is transmitting. Now adjust the background noise out by turning the squelch control clockwise. Some handhelds have a menu squelch setting, so play around until you get to the point where the background noise is just gone when no one is on the air. **ANSWER D.**

T2B03 Which of the following describes the muting of receiver audio controlled solely by the presence or absence of an RF signal?

A. Tone squelch.
B. Carrier squelch.
C. CTCSS.
D. Modulated carrier.

The most basic carrier squelch circuit in your new radio will nicely hush background static when the repeater or local signal stops transmitting. It would drive you nuts to have to listen to background hash when the repeater is not in use! The squelch circuit usually is preset by either a handheld top knob, or a squelch level setting as a handheld or mobile radio menu item. Sometimes a carrier squelch circuit has been pre-set, so you won't need to do a thing other than turn up the volume, and wait for a signal to come in on your handheld or mobile radio, overriding the ***carrier squelch circuit***. **ANSWER B.**

T2B01 What is the term used to describe an amateur station that is transmitting and receiving on the same frequency?

A. Full duplex communication.
B. Diplex communication.
C. Simplex communication.
D. Half duplex communication.

Simplex means ***same frequency***. Operate simplex on VHF or UHF when the other station is within a few miles of your station. The opposite of simplex is duplex, a type of repeater operation. A great way to get started with your brand new call sign is to transmit on 146.520 MHz, simplex. This is the national calling channel, and a great place to meet nearby new ham friends. On 70 cm, try simplex on 446.000 MHz. You will be surprised by how far your signal may go if you get up on a hill or high up on top of a building. **ANSWER C.**

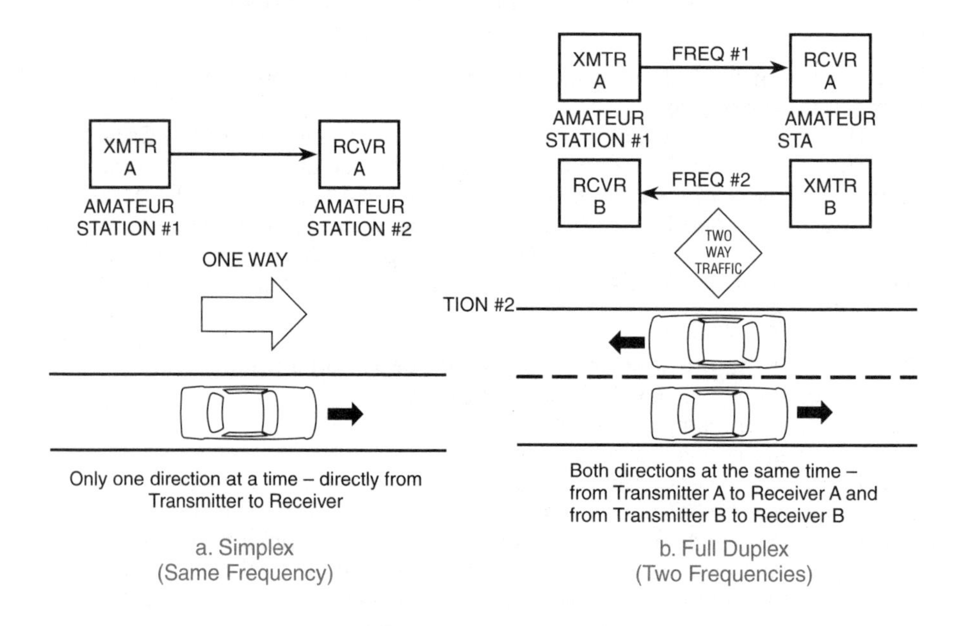

Simplex and Duplex Communications

T2A06 What must an amateur operator do when making on-air transmissions to test equipment or antennas?

A. Properly identify the transmitting station.
B. Make test transmissions only after 10:00 p.m. local time.
C. Notify the FCC of the test transmission.
D. State the purpose of the test during the test procedure.

Even a short ***test transmission*** to check out your new antenna requires ***station identification***. Try to do your antenna testing on a simplex frequency to avoid tying up a repeater unnecessarily. **ANSWER A.**

Ham Hint: *Ham radio operators enjoy tinkering with their equipment. If you're at your workbench testing the quality of your transmissions by sending a signal into a non-radiating dummy load, there is still plenty of signal for you to monitor with another receiver. Try to avoid any continuous testing when hooked up to an outside antenna, unless you are testing the antenna itself.*

T2A07 Which of the following is true when making a test transmission?

A. Station identification is not required if the transmission is less than 15 seconds.
B. Station identification is not required if the transmission is less than 1 watt.
C. Station identification is required only if your station can be heard.
D. Station identification is required at least every ten minutes during the test and at the end.

Ham radio operators enjoy precisely tuning their antenna systems for peak performance. This means actually radiating a signal on the airwaves, and looking at in-line antenna test equipment for the best possible signal transfer. The good ham always looks for an unused frequency, and runs the minimum power necessary to get an adequate reading on their in-line antenna test meter. The ham should identify regularly, and must always ***identify*** at least ***every 10 minutes and when they sign off*** using their call sign after conducting their test. **ANSWER D.**

T2A08 What is the meaning of the procedural signal "CQ"?

A. Call on the quarter hour.
B. A new antenna is being tested (no station should answer).
C. Only the called station should transmit.
D. Calling any station.

The 2 letters ***"CQ" mean "calling any station,"*** and we use this on all worldwide bands and weak signal calls over VHF and UHF frequencies. But CQ is not ever used when operating on FM repeater and simplex frequencies because the presence of your FM carrier is strong enough to let everyone else know you are on the air. Instead of calling "CQ" over a repeater, you simply announce your call letters and indicate you are monitoring for a call. And if you're on the air for the very first time, tell them you are a Gordo grad, and that may be all that is necessary to bring back plenty of responses from the ham community welcoming you to the exciting airwaves! Try this: "This is (your call sign repeated twice, phonetically) on the air for the first time, a friend of Gordon West, looking for my very first contact over ham radio. Over." This will certainly get attention, and likely you'll get plenty of radio calls. Always be sure to say your call sign slowly, using the phonetic alphabet. **ANSWER D.**

Ham Hint: Getting On The Air!

As you prepare to pass the exam, get set to join a fraternity of fellow ham radio operators. In less than a month, studying one section a day, you will be prepared to find your local examination session: Visit www.W5YI.org or www.ARRL.org/examsearch to find an exam site near you.

If you get stuck in your studies, call Gordo direct (Monday through Thursday, 10 am to 4 pm California time) at 714-549-5000. Gordo will be your personal Elmer.

When you pass the exam, ask your VE team for local club information. Most clubs operate on repeater channels on 2 meters and 70 cm. JOIN A HAM CLUB! The club members will give you lots of help learning the amateur radio ropes. To find a ham club in your area, visit www.ARRL.org/clubsearch.

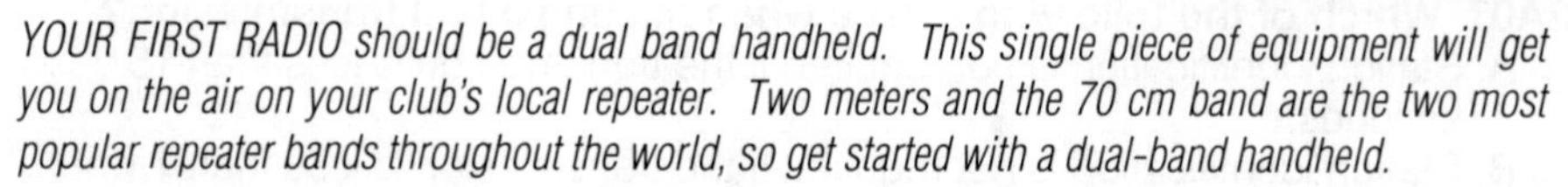

YOUR FIRST RADIO should be a dual band handheld. This single piece of equipment will get you on the air on your club's local repeater. Two meters and the 70 cm band are the two most popular repeater bands throughout the world, so get started with a dual-band handheld.

Some good accessories for your 2m/440 MHz handheld are an alkaline battery tray, extra-long rubber antenna, and a combination speaker/microphone. For mobile use, get the 12 volt DC car adapter, light-weight mobile magnetic mount antenna, the mobile antenna adapter, and a hands-free headset for use while driving.

OK, you got your radio and all the goodies. What's next? A very few ham radio sellers may pre-program local repeater channels in the area where you live and work. Purchasing the Repeater Directory will help, and getting your local club member to "clone" or pre-program channels into your handheld can get you going in a hurry!

If you end up with a dual-band handheld with nothing pre-programmed, just remember R-O-O-T. Let's get to the root of the programming function:

First, dial in REPEATER OUTPUT

Next, check that the OFFSET automatically comes up on the screen, or enter the offset MINUS or PLUS

Use the Repeater directory to look up TONE. Tone must first be switched on to ENCODE, and then the specific TONE CODE must be entered.

Now, try to bring up the repeater by pushing PUSH-TO-TALK, and listen for the beep. Always give your call sign during this test transmission. Got Beep? Now, MEMORIZE this in an open memory channel.

Trust me, get your radio cloned by a local club member or ask a pal to load in some local hot repeater channels.

T2A05 What should you transmit when responding to a call of CQ?

A. CQ followed by the other station's call sign.
B. Your call sign followed by the other station's call sign.
C. The other station's call sign followed by your call sign.
D. A signal report followed by your call sign.

What a great day on 6 meters! Signals are coming in via skywaves from over 1,000 miles away. And up at 50.140, here is a young lady calling CQ, and then she stands by for a return call. It's your turn to transmit! IMMEDIATELY key your microphone, ***say HER call sign once, and then give YOUR call sign phonetically a couple of times***. Release your push-to-talk button, and chances are she will return YOUR call sign and your conversation begins on 6 meters skywave. **ANSWER C.**

T2A04 What is an appropriate way to call another station on a repeater if you know the other station's call sign?

A. Say "break, break" then say the station's call sign.
B. Say the station's call sign then identify with your call sign.
C. Say "CQ" three times then the other station's call sign.
D. Wait for the station to call "CQ" then answer it.

Before transmitting on any frequency, be sure to listen for a few seconds to insure the channel is clear. Then depress the microphone push-to-talk button and ***say the call sign of the station you are wishing to hook up with, followed by your call sign***, phonetically, and the optional word "over." **ANSWER B.**

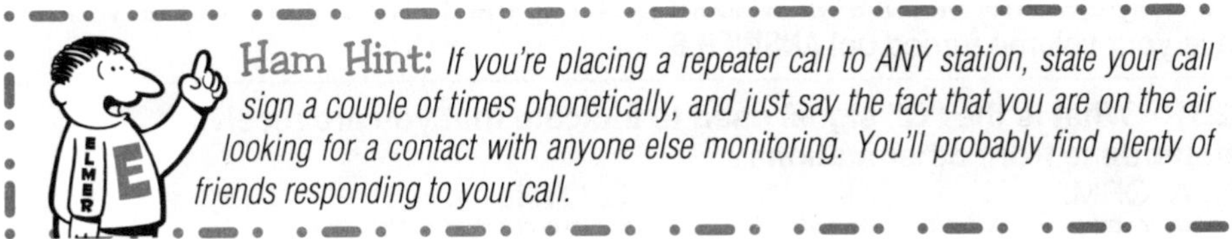

Ham Hint: *If you're placing a repeater call to ANY station, state your call sign a couple of times phonetically, and just say the fact that you are on the air looking for a contact with anyone else monitoring. You'll probably find plenty of friends responding to your call.*

T3A01 What should you do if another operator reports that your station's 2 meter signals were strong just a moment ago, but now they are weak or distorted?

A. Change the batteries in your radio to a different type.
B. Turn on the CTCSS tone.
C. Ask the other operator to adjust his squelch control.
D. Try moving a few feet, as random reflections may be causing multi-path distortion.

Move one step to the left or right and presto, you are now loud and clear! If you are transmitting with your handheld inside a building, radio waves "illuminate" the inside of a building much like sunlight through Venetian blinds. You will find "HOT spots" and "NOT spots." The Not spots are much like sunshine shadows, and the hot spots are where your signal is getting outside quite nicely. These locations usually won't change to a distant repeater, so remember where your in-house hot spots are, and don't expect to get much when transmitting down in the basement! **ANSWER D.**

T3A06 What term is commonly used to describe the rapid fluttering sound sometimes heard from mobile stations that are moving while transmitting?

A. Flip-flopping.
B. Picket fencing.
C. Frequency shifting.
D. Pulsing.

When mobile stations running either a handheld to an outside antenna, or a 50-watt mobile to an outside antenna, get to the end of their line-of-sight range to a repeater, the signal will begin to ***flutter rapidly***, caused by roadway signs and guardrails adding and subtracting to the ultimate signal strength. In the two-way radio business, we call this ***"picket fencing."*** See? A new term for your ham radio vocabulary! **ANSWER B.**

T2B08 What is the proper course of action if your station's transmission unintentionally interferes with another station?

A. Rotate your antenna slightly.
B. Properly identify your transmission and move to a different frequency.
C. Increase power.
D. Change antenna polarization.

You are driving down the road and decide you want to start a conversation with someone on the repeater. You turn on the radio, dial in the appropriate repeater channel, and hear nothing but silence. You surmise no one is using the repeater, so you give your call sign and tell the world you're traveling down I-95 and are looking for a conversation. You glance down and see your volume is all the way down, and when you turn it up, you discover you transmitted right on top of an ongoing conversation. What should you do? The polite ham will ***wait for a break in the conversation***, then ***announce your own call sign*** and a quick apology for accidentally transmitting on top of an ongoing conversation. We all have done it, and likely the other stations will welcome you into the conversation now that you have your volume turned up! **ANSWER B.**

T2B10 What is the "Q" signal used to indicate that you are receiving interference from other stations?

A. QRM.
B. QRN.
C. QTH.
D. QSB.

When skywaves bounce off of the ionosphere on the 6-meter band, you'll sometimes hear several stations returning your CQ (calling any station) call, all transmitting at the same time. This is called ***"QRM," the Q-code for many other stations interfering*** with each other. **ANSWER A.**

T2B11 What is the "Q" signal used to indicate that you are changing frequency?

A. QRU.
B. QSY.
C. QSL.
D. QRZ.

If you find that YOUR station is accidentally causing QRM to an ongoing conversation, it is time to ***QSY – change to another frequency***. **ANSWER B.**

T8C03 What popular operating activity involves contacting as many stations as possible during a specified period of time?

A. Contesting.
B. Net operations.
C. Public service events.
D. Simulated emergency exercises.

While you won't hear a lot of this on your 2-meter/70-centimeter (440) handheld, you WILL hear this on 6 meters and 2-meter single sideband – ***CONTESTING***. It usually takes place every couple of months over a single weekend. The idea is to contact as many other stations and exchange specific station details, as if you were handling an emergency message. Contesting is a great way to double-check the performance of your radio system, and it assists you in preparing for an emergency when you may need to contact as many stations as possible. **ANSWER A.**

POPULAR Q SIGNALS

Given below are a number of Q signals whose meanings most often need to be expressed with brevity and clarity in amateur work. (Q abbreviations take the form of questions only when each is sent followed by a question mark.)

QRG Will you tell me my exact frequency (or that of _____)? Your exact frequency (or that of _____) is _____ kHz.
QRH Does my frequency vary? Your frequency varies.
QRI How is the tone of my transmission? The tone of your transmission is _____ (1. Good; 2. Variable; 3. Bad).
QRJ Are you receiving me badly? I cannot receive you. Your signals are too weak.
QRK What is the intelligibility of my signals (or those of _____)? The intelligibility of your signals (or those of _____) is _____ (1. Bad; 2. Poor; 3. Fair; 4. Good; 5. Excellent).
QRL Are you busy? I am busy (or I am busy with _____). Please do not interfere.
QRM Is my transmission being interfered with? Your transmission is being interfered with _____ (1. Nil; 2. Slightly; 3. Moderately; 4. Severely; 5. Extremely).
QRN Are you troubled by static? I am troubled by static _____ (1-5 as under QRM).
QRO Shall I increase power? Increase power.
QRP Shall I decrease power? Decrease power.
QRQ Shall I send faster? Send faster (_____ WPM).
QRS Shall I send more slowly? Send more slowly (_____WPM).
QRT Shall I stop sending? Stop sending.
QRU Have you anything for me? I have nothing for you.
QRV Are you ready? I am ready.
QRW Shall I inform _____ that you are calling on _____ kHz? Please inform _____ that I am calling on _____ kHz.
QRX When will you call me again? I will call you again at _____ hours (on _____ kHz).
QRY What is my turn? Your turn is numbered _____.
QRZ Who is calling me? You are being called by _____ (on _____ kHz).
QSA What is the strength of my signals (or those of _____)? The strength of your signals (or those of _____) is _____ (1. Scarcely perceptible; 2. Weak; 3. Fairly good; 4. Good; 5. Very good).
QSB Are my signals fading? Your signals are fading.
QSD Is my keying defective? Your keying is defective.
QSG Shall I send _____ messages at a time? Send _____ messages at a time.
QSK Can you hear me between your signals and if so can I break in on your transmission? I can hear you between my signals; break in on my transmission.
QSL Can you acknowledge receipt? I am acknowledging receipt.
QSM Shall I repeat the last message which I sent you, or some previous message? Repeat the last message which you sent me [or message(s) number(s) _____].
QSN Did you hear me (or _____) on _____ kHz? I heard you (or _____) on _____ kHz.
QSO Can you communicate with _____ direct or by relay? I can communicate with _____ direct (or by relay through _____).
QSP Will you relay to _____? I will relay to _____.
QST General call preceding a message addressed to all amateurs and ARRL members. This is in effect "CQ ARRL."
QSU Shall I send or reply on this frequency (or on _____ kHz)?
QSW Will you send on this frequency (or on _____ kHz)? I am going to send on this frequency (or on _____ kHz).
QSX Will you listen to _____ on _____ kHz? I am listening to _____ on _____ kHz.
QSY Shall I change to transmission on another frequency? Change to transmission on another frequency (or on _____ kHz).
QSZ Shall I send each word or group more than once? Send each word or group twice (or _____ times).
QTA Shall I cancel message number _____? Cancel message number _____.
QTB Do you agree with my counting of words? I do not agree with your counting of words. I will repeat the first letter or digit of each word or group.
QTC How many messages have you to send? I have messages for you (or for _____).
QTH What is your location? My location is _____.
QTR What is the correct time? The time is _____.

Source: ARRL

T8C04 Which of the following is good procedure when contacting another station in a radio contest?

A. Be sure to sign only the last two letters of your call if there is a pileup calling the station.
B. Work the station twice to be sure that you are in his log.
C. Send only the minimum information needed for proper identification and the contest exchange.
D. All of these choices are correct.

During Field Day, as well as special weekends during the year, our ham bands become more populated with contesters. We try to work as many other stations as we can in many different geographic areas. Many contesters are dead serious about winning, so they don't have time for long-winded pleasantries during the contest exchange. You will ***only transmit your call sign and the information they need***, and save chitchat for another weekend! These contesters need to make the exchange fast and move on quickly. **ANSWER C.**

T8C05 What is a grid locator?

A. A letter-number designator assigned to a geographic location.
B. A letter-number designator assigned to an azimuth and elevation.
C. An instrument for neutralizing a final amplifier.
D. An instrument for radio direction finding.

When you get active on VHF weak-signal work using a multi-mode radio for single sideband, you'll hear the weak-signal operators saying their ***location as a grid square***. Grid squares are 2-letter by 2-number designators based on 1 degree latitude by 2 degrees longitude. Grid square maps are a handy operating aid when the band opens on 6 meters! My location is DM13, and last night a chap broke in and gave me a signal report from EL95, the tip of Florida! Imagine my surprise, but not that unusual for the 6-meter magic band that you'll hear on the audio CD accompanying this book. So check out your grid square map for weak-signal VHF/UHF operating – more than likely, another station is going to ask, "What's your grid?" When you pass your test and receive your new call letters, be sure to let me know. I'll send you a free graduation package including a large grid square map of the USA. Read about my free graduation package on page 187. I want to hear from all of you when you pass the exam and send you some valuable operating aids along with your graduation certificate, which I will sign personally. **ANSWER A.**

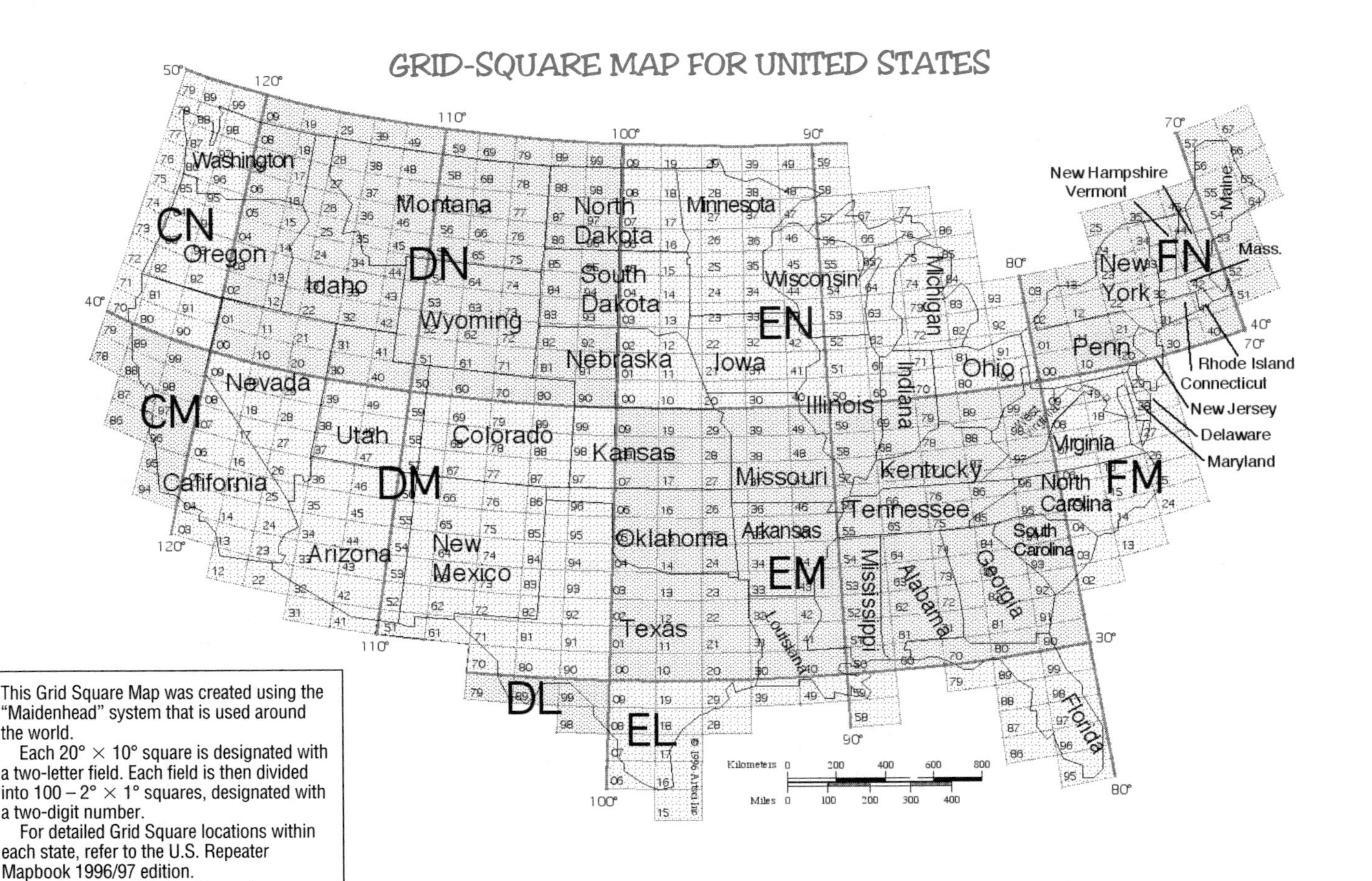

This Grid Square Map was created using the "Maidenhead" system that is used around the world.

Each 20° × 10° square is designated with a two-letter field. Each field is then divided into 100 – 2° × 1° squares, designated with a two-digit number.

For detailed Grid Square locations within each state, refer to the U.S. Repeater Mapbook 1996/97 edition.

http://home.earthlink.net/~artsci

T3A02 Why are UHF signals often more effective from inside buildings than VHF signals?

A. VHF signals lose power faster over distance.
B. The shorter wavelength allows them to more easily penetrate the structure of buildings.
C. This is incorrect; VHF works better than UHF inside buildings.
D. UHF antennas are more efficient than VHF antennas.

Recall that I recommend a dual-band handheld as your first radio. If it has more than 2 bands, fine, but the main 2 bands you want will be 2 meters and 70 centimeters (the 440 MHz band). Signals on the 70-cm, 440 MHz band, are ***short enough in wavelength that they*** love bouncing around inside buildings and ***penetrating walls***. Elevators, too! Even bank vaults if they close the door. So, just because there is a lot of metal around you that you think would shield your signal from the outside world, presto, your UHF signals may make it out. **ANSWER B.**

Ham Hint: *By the way, standing next to one of those big windows that look on the outside world does not necessarily give you a better shot to the distant repeater. Many times those windows have a metallic ultraviolet shield built in, and they actually block the radio signals rather than pass them. So walk around your high-rise office and find a "hot spot" to the distant repeater, and it might not be by your big window, but rather in an area of the building where the signals can make it out to the outside world. If you always stand in the "hot spot," you'll usually always reach that distant repeater.*

Website Resources

▾ IF YOU'RE LOOKING FOR	▾ THEN VISIT
Operating Hints for Icom Radios	www.icomelmer.com
Ham Clubs Near You!	www/ARRL.org/FandES/Field/Club/Clubsearch/phtml
Just For Ladies	www.YLRL.org

Repeaters

T1F09 What type of amateur station simultaneously retransmits the signal of another amateur station on a different channel or channels?

A. Beacon station.
B. Earth station.
C. Repeater station.
D. Message forwarding station.

The device that retransmits amateur radio signals within a specific ham band is called a repeater. When you get your new Technician Class license, first do about one week of monitoring – without transmitting – listening to repeater communications. This will give you a good idea on what the proper operating procedures are for that local repeater frequency. I also recommend you join the local repeater club and let club members help you program your new radio equipment for some of the local ***repeater*** frequencies. [97.3(a)(39)] **ANSWER C.**

Before you press the PTT switch, LISTEN to make sure the frequency is clear.

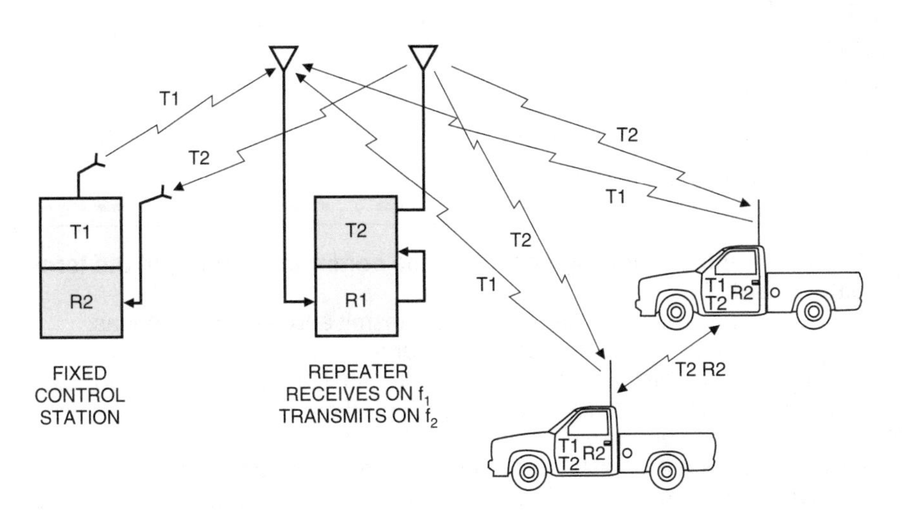

Repeater

Source: *Mobile 2-Way Radio Communications,* G. West, Copyright ©1992 Master Publishing, Inc., Niles, IL

T1D07 When is an amateur station authorized to automatically retransmit the radio signals of other amateur stations?

A. When the signals are from an auxiliary, beacon, or Earth station.
B. When the signals are from an auxiliary, repeater, or space station.
C. When the signals are from a beacon, repeater, or space station.
D. When the signals are from an Earth, repeater, or space station.

Up on a mountaintop, ham operators have established ***auxiliary and repeater stations*** which automatically re-transmit your signal to other ham stations. Out in space, ***space stations*** may also incorporate automatic equipment to re-transmit the radio signal of ham stations. If you are into computers, imagine sending "Hello World," in data, along with your call sign through the International Space Station, and getting a response a few hours later from the other side of the world! [97.113(f)] **ANSWER B.**

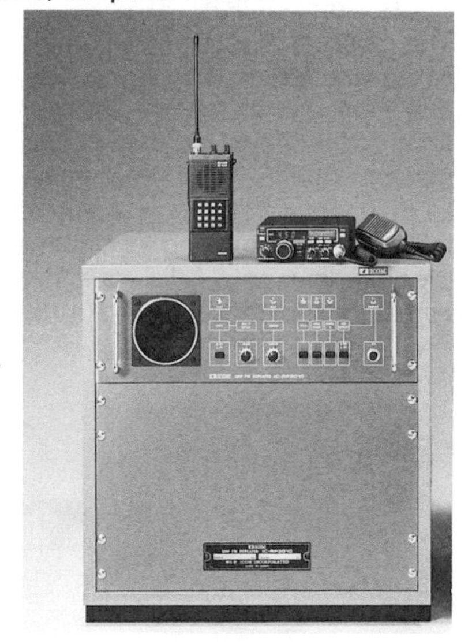

A Repeater

Ham Hint: *Even base stations are permitted to use repeaters for added communications distance. There is usually no charge for joining a repeater group. Some repeaters have autopatch, and those repeaters may require special access codes and financial support.*

Try these sites:
www.winsystem.org
www.wa6twf.com

T4B11 Which of the following describes the common meaning of the term "repeater offset"?

A. The distance between the repeater's transmit and receive antennas.
B. The time delay before the repeater timer resets.
C. The difference between the repeater's transmit and receive frequencies.
D. The maximum frequency deviation permitted on the repeater's input signal.

When you purchase your new dual-band handheld to get started as a new Technician Class operator, ask the dealer to preprogram several local, popular repeater channels in your area. They might do this on a computer, selecting the repeater's output, and the repeater offset – ***the difference between the repeater's output and*** where the repeater receives your ***input transmit frequencies***. You also may need a required CTCSS sub-audible tone to activate a repeater to receive your transmit signal. **ANSWER C.**

T2A01 What is the most common repeater frequency offset in the 2 meter band?

A. Plus 500 kHz.
B. Plus or minus 600 kHz.
C. Minus 500 kHz.
D. Only plus 600 kHz.

I am happy to report that most of your new 2 meter/440 MHz dual-band ham radio equipment has automatic repeater offset capability. This means, on the ***2 meter*** band, your transmitter will ***offset plus or minus 600 kHz***. It could go either way, depending on the band plan. Plus or minus. Only one answer with both plus or minus. **ANSWER B.**

T2A03 What is a common repeater frequency offset in the 70 cm band?

A. Plus or minus 5 MHz.
B. Plus or minus 600 kHz.
C. Minus 600 kHz.
D. Plus 600 kHz.

Happy to report that modern, dual-band ham radio equipment offers automatic offsets on the ***70 cm***, 440 MHz, band, too. We use a ***5 MHz offset***, plus or minus, on the 70 cm band. **ANSWER A.**

T2B04 Which of the following common problems might cause you to be able to hear but not access a repeater even when transmitting with the proper offset?

A. The repeater receiver requires audio tone burst for access.
B. The repeater receiver requires a CTCSS tone for access.
C. The repeater receiver may require a DCS tone sequence for access.
D. All of these choices are correct.

You just brought home your new dual-band, handheld ham radio, but forgot to ask the seller to preprogram some of your local channels. Always get your new transceiver preprogrammed for a few of your local repeater frequencies. It will prevent initial disappointment because there's a lot to know at first. And today's frustration is that you've tuned into a local powerful repeater, just 1 mile away, but no matter what you do, no one hears you respond when they call you during the "net." Most likely, you missed the correct offset, or the repeater may require a tone burst that you didn't know about, or, more than likely, the repeater requires a specific CTCSS tone in order for your signal to pass through. Which tone is right? Or maybe the repeater requires a digital coded squelch tone sequence for access, and you don't have a clue. ***All of these*** could spell frustration, unless you get your radio preprogrammed by the dealer, or your local repeater club. Once you have the right tones and offset in, you'll be ON! **ANSWER D.**

Ham Hint: *A handy little booklet called a "Repeater Guide" or "Repeater Atlas" will help you identify the active repeaters in your local area, or active repeaters in a distant city where you plan to travel. There also are software repeater guides that will let you compute your upcoming road trip and all of the local repeaters along the way will pop up on the screen, along with their input and output frequencies, sub-audible tone requirements, and a color coverage map showing the approximate range of that repeater along your route. Avoid programming your handheld while driving! Get those repeater channels into memory before you head out, and then enjoy ham radio on the road with your new dual-band handheld.*

T2B02 What is the term used to describe the use of a sub-audible tone transmitted with normal voice audio to open the squelch of a receiver?

A. Carrier squelch. C. DTMF.
B. Tone burst. D. CTCSS.

Every mountain top and skyscraper probably has a few ham repeaters atop a small tower. These repeaters are hearing so many signals coming in, including interfering signals, that they need a way of not accidentally self-triggering and turning on when it wasn't the real signal. What makes a signal into a ham repeater REAL? The repeater may employ ***CTCSS tone*** decode, and it will take the ham out there in radioland to ENCODE a specific subaudible tone which causes the repeater receiver to accept the signal. **ANSWER D.**

Ham Hint: You Need The Tone, Too!

Just as important as the correct repeater split frequency is the correct tone. Tone? Your new dual band handheld radio or mobile radio has a tone encode and decode feature called Continuous Tone Coded Squelch, CTCSS, for short, or "tone" for real short. You must select the correct tone for the repeater you have chosen to transmit on. There are 40+ tone possibilities, and the repeater guide will tell you what tone to ENCODE in your radio. A little "t", or "enc" will show up on your screen, indicating you are transmitting a slight "hum" to activate that repeater. The tone will not allow access if the repeater is considered "closed." You need the correct signal activate the closed repeater system. Your radios have the tone circuit already built in, so it is up to you, or the ham programming your radio, to encode the correct tone for that particular repeater. And if the repeater IS closed, first check with the repeater owner to obtain the correct tone to access the system.

Remember, support your local repeaters, open or closed!

EIA Standard Subaudible CTCSS (PL) Tone Frequencies

Freq.	Tone No.	Tone Code	Freq.	Tone No.	Tone Code	Freq.	Tone No.	Tone Code
67.0	01	XZ	110.9	15	2Z	179.9	29	6B
71.9	02	XA	114.8	16	2A	186.2	30	7Z
74.4	03	WA	118.8	17	2B	192.8	31	7A
77.0	04	XB	123.0	18	3Z	203.5	32	M1
79.7	05	SP	127.3	19	3A	206.5		8Z
82.5	06	YZ	131.8	20	3B	210.7	33	M2
85.4	07	YA	136.5	21	4Z	218.8	34	M3
88.5	08	YB	141.3	22	4A	225.7	35	M4
91.5	09	ZZ	146.2	23	4B	229.2		9Z
94.8	10	ZA	151.4	24	5Z	233.6	36	
97.4	11	ZB	156.7	25	5A	241.8		M5
100.0	12	1Z	162.2	26	5B	250.3		M6
103.5	13	1A	167.9	27	6Z	256.3		M7
107.2	14	1B	173.8	28	6A			

T2A09 What brief statement is often used in place of "CQ" to indicate that you are listening on a repeater?

A. Say "Hello test" followed by your call sign.
B. Say your call sign.
C. Say the repeater call sign followed by your call sign.
D. Say the letters "QSY" followed by your call sign.

Simply ***say your call sign*** and announce that you are monitoring for any call. **ANSWER B.**

Ham Hint: *Okay, you have your brand new dual-band handheld, and it's tied into your hidden attic antenna. Your batteries are all charged up, and the local radio dealer memorized a wonderful repeater that you've been listening to all evening long. Now it's time for YOU to make your first transmission.*

First, listen for a couple of minutes to make sure the repeater is not in use. Next, momentarily adjust the squelch so you get background noise, and then adjust the volume to about mid-scale. Now re-adjust the squelch to block background noise. Then press the push-to-talk for approximately 2 seconds, let go, and listen for the repeater to go BEEP. This is the repeater courtesy tone, and most repeaters have a tone that comes on after you release the microphone button to signal the end of that particular transmission. A few repeaters may just show up as a strong single on your handheld signal-strength meter, and then silently click off.

Whether you reached a repeater or not, you MUST give your call sign – so press the push-to-talk button, wait about 1 second, and then clearly state your call letters phonetically: "Kilo zulu six hotel alpha mike, on the air for the first time, brand new ham, listening. Over." Now release the push-to-talk button.

You should hear a beep to confirm your signal was indeed passed through the repeater, and a few seconds later you will probably hear someone calling your call sign and then giving their call sign. If you can, try to write down their call sign.

Now it's your turn to talk. Wait for the beep! You'll press the push-to-talk button, and nothing will come out of your mouth! After all, this is your first transmission, and stage fright is very common. Tell them your first name, where you are located, the fact that this is your first transmission and you're scared to pieces, and maybe find out where the local ham radio club meets because your instructor, Gordo, always said to join a local ham club. Now say over, and release the push-to-talk button.

After that, you'll be rolling with ham radio. Be sure to give your call sign every 10 minutes and when you sign off.

T1A08 Which of the following entities recommends transmit/receive channels and other parameters for auxiliary and repeater stations?

A. Frequency Spectrum Manager.
B. Frequency Coordinator.
C. FCC Regional Field Office.
D. International Telecommunications Union.

Frequency coordination for VHF and UHF band plans is developed by ***regional frequency coordinators***. It is a huge job because seasoned hams all want their own repeater frequency pairs. Now add requests for simplex coordination for voice-over-Internet systems, and you will see that frequency coordination in any local area must balance the needs of all ham radio operators. When you purchase your new two-band VHF/UHF handheld, be sure to buy a USA repeater atlas. This way, when you see your family and friends in San Diego, Seattle, Miami and Connecticut, you'll know what frequencies and what tones to use for the local repeaters. The "locals" will have fun working you on the airwaves as an out of town guest, and even though you may be driving in a strange area, you will have plenty of backup from all your new ham radio friends around you using their local repeater. [97.3(a)(22)]
ANSWER B.

Ham Hint: *Repeater directories publish the repeater frequencies by output. The plus (+) or minus (-) indicates the input "split" that you dial in on your VHF or UHF ham set. A plus (+) indicates a higher input and a minus (-) indicates a lower input. When you start to transmit, your transmitter should automatically go to the proper input frequency. Some repeaters also require a sub-audible tone as part of your input transmission. Ask the local operators how to engage the tone signal on your ham radio set.*

T1A09 Who selects a frequency coordinator?

A. The FCC Office of Spectrum Management and Coordination Policy.
B. The local chapter of the Office of National Council of Independent Frequency Coordinators.
C. Amateur operators in a local or regional area whose stations are eligible to be auxiliary or repeater stations.
D. FCC Regional Field Office.

The first thing you'll want to do when your new ham ticket arrives (and I know you're going to pass the test on the first try) is to get on the air. Start off with a dual-band handheld and enjoy all of the excitement on 2 meters and the 440 MHz band. Here is where you will find all of your local repeaters, and in a few more pages, you will learn all that repeaters can do to send your voice throughout your area and, many times, throughout the country and around the World! In urban areas, repeaters on building tops and mountain peaks will fill your radio dial with fascinating conversations. Each repeater is assigned a specific set of operating frequencies, so they don't interfere with each other – like designated parking stalls. Frequency coordinators are fellow ham operators in a local or regional area, voted in to their positions by ***fellow hams who are active repeater station owners or auxiliary station operators***. This is a volunteer job. Frequency coordination keeps our ham radio frequencies and bands clear of interference. [97.3(a)(22)]
ANSWER C.

T1A11 Which of the following stations transmits signals over the air from a remote receive site to a repeater for retransmission?

A. Beacon station.
B. Relay station.
C. Auxiliary station.
D. Message forwarding station.

A small handheld radio can sound like a powerful home base station, thanks to repeaters. You transmit "Hello World," and a nearly simultaneous repeat of your voice comes out from the distant mountaintop. Maybe you are in the next town over, and that distant repeater 90 miles away, running 100 watts, isn't coming in as strong as before. Good news – ham repeater owners may have set up a repeater ***auxiliary station*** to capture local, next-town-over signals loud and clear for re-transmission to the main repeater. Auxiliary stations are remote receiving sites. Some repeaters have as many as 6 auxiliary stations, controlled by a voting system to pick out the best auxiliary input station signal. The term "voting" describes specialized repeaters with multiple area inputs. The inputs are simultaneously microwaved to the repeater receiver voting system. If the voting system finds a stronger signal on one of the inputs, that is the signal that is delivered to the repeater output. Since all inputs are on microwave frequencies back to the main repeater, the voting system is supported by the term "auxiliary station." So now you know how that distant repeater heard your tiny handheld signal, so many miles away! [97.3(a)(7)] **ANSWER C.**

T1F05 What method of call sign identification is required for a repeater station retransmitting phone signals?

A. Send the call sign followed by the indicator RPT.
B. Send the call sign using CW or phone emission.
C. Send the call sign followed by the indicator R.
D. Send the call sign using only phone emission.

Some repeaters have a young lady that greets you with a repeater call sign, the time, and have a nice day! Some may even give you the local temperature at the repeater site. If you are sending and receiving slow-scan and fast-scan pictures and television signals, you can ***run a video clip*** that will show your ***call sign***, and this meets the rules. If it's not practical to run a video clip, every 10 minutes pan your camera to your cap that has your call sign on it, or speak into your ATV microphone with your FCC-assigned call letters. Of course, ***Morse code (CW) is always allowed as an identification method***, but don't exceed 20-wpm because the FCC says so and that is about the fastest that I can copy myself! [97.119(b)] **ANSWER B.**

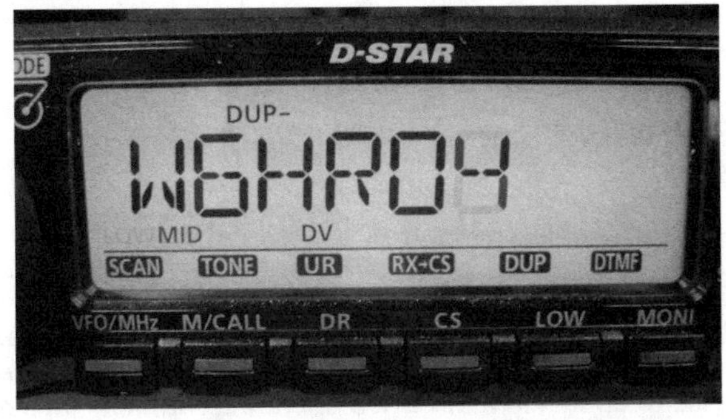

Repeaters can identify with a voice message announcing their call sign, or use Morse code to send out their station call letters.

Emergency!!!

T2C06 Which of the following is common practice during net operations to get the immediate attention of the net control station when reporting an emergency?

A. Repeat the words SOS three times followed by the call sign of the reporting station.
B. Press the push-to-talk button three times.
C. Begin your transmission with "Priority" or "Emergency" followed by your call sign.
D. Play a pre-recorded emergency alert tone followed by your call sign.

You are out at sea listening to your local maritime net with fellow sailors describing how warm the water is in the Bahamas. You glance down and notice that your ship's floorboards are floating around in the cabin, and the salt water is rising above your knees. You're sinking! Grab the microphone and interrupt the net with ***"priority" or "emergency" followed by your call sign***. This will get the immediate attention of the net control station to report your emergency. **ANSWER C.**

Another way to interrupt a conversation to signal a distress call is to say the word "BREAK" several times to indicate a priority or emergency distress call. Keep this in mind when operating routinely on a repeater – don't say the word "break" unless it's an emergency or something very, very important—like being stuck in the mud surrounded by alligators!

T2C09 When may an amateur station use any means of radio communications at its disposal for essential communications in connection with immediate safety of human life and protection of property?

A. Only when FEMA authorizes it by declaring an emergency.
B. When normal communications systems are not available.
C. Only when RACES authorizes it by declaring an emergency.
D. Only when authorized by the local MARS program director.

In an ***emergency, anything goes!*** But remember, using a ham radio on an alternate radio service like the marine radio service requires a declaration of an emergency, with a life-and-death situation on hand. When sailors head out to remote areas well beyond reliable radio range to other amateur operators, in a life-and-death situation, they could switch to the United States Coast Guard safety channel to request assistance and give their location, preferably as latitude and longitude. [97.403] **ANSWER B.**

Amateur radio operators are well known for their volunteer assistance in emergencies—from local problems to national disasters like 9/11 and Hurricane Katrina

Ham Hint: *Here's an important exception to the fundamental rule that you can use any radio on any frequency to summon help in an emergency: Avoid contacting a radio service that prohibits any radio call coming in from unknown units. This would include secure military nets, law enforcement radio service, fire radio service, FBI, and other agencies that ONLY communicate among themselves. Using your ham radio for out-of-band transmissions to local police, fire, and state agencies could cost you your ham license when proven you had other radio services, like the U.S. Coast Guard, that you could call who stand by for incoming emergency radio traffic.*

T2C07 What should you do to minimize disruptions to an emergency traffic net once you have checked in?

A. Whenever the net frequency is quiet, announce your call sign and location.
B. Move 5 kHz away from the net's frequency and use high power to ask other hams to keep clear of the net frequency.
C. Do not transmit on the net frequency until asked to do so by the net control station.
D. Wait until the net frequency is quiet, then ask for any emergency traffic for your area.

Check in just once, and ***don't transmit again until directed to do so by the net control operator***. **ANSWER C.**

Ham Hint: *During an emergency net, the net control station needs all incoming transmissions to be brief and contain only the information they ask for. Once you pass that information, go into the listen mode. In a widespread emergency, an undisciplined check-in where the ham goes on for 3 minutes to describe all the training he has received over the years in emergency preparedness simply clogs the network.*

T2C01 What set of rules applies to proper operation of your station when using amateur radio at the request of public service officials?

A. RACES Rules.
B. ARES Rules.
C. FCC Rules.
D. FEMA Rules.

Your amateur license is governed by the Federal Communications Commission. You are bound by ***FCC rules***, so any request from the FBI, FEMA, or any other Federal agency does not relieve you from obeying FCC rules. [97.103(a)] **ANSWER C.**

T2C05 What is the Radio Amateur Civil Emergency Service?

A. An emergency radio service organized by amateur operators.
B. A radio service using amateur stations for emergency management or civil defense communications.
C. A radio service organized to provide communications at civic events.
D. A radio service organized by amateur operators to assist non-military persons.

RACES stands for Radio Amateur Civil Emergency Service. It is a division of the civil defense organization that ***uses ham stations for emergency and CD communications***. You must be registered to take part in RACES drills. **ANSWER B.**

RACES Logo.

T2C04 What do RACES and ARES have in common?

A. They represent the two largest ham clubs in the United States.
B. Both organizations broadcast road and weather traffic information.
C. Neither may handle emergency traffic supporting public service agencies.
D. Both organizations may provide communications during emergencies.

Providing emergency communications as an active ham member of ARES or RACES requires a lot more than having your ham radio all set to go and some flashy magnetic signs for your car. Regular training is the absolute key to an effective emergency communications team. **ANSWER D.**

In an emergency, authorized hams participating in a RACES organization may communicate from a police helicopter.

Ham Hint: *As an ARES or RACES member, you can expect to be on the air at least once or twice a week on your specific net time. Once a month you will attend a local meeting, and regularly you will accrue additional training to make you a better emergency responder. You will have a distinctive uniform, a set callout plan when you are informed that a major emergency has occurred nearby, and a ham radio "grab-and-go" kit that will keep you on the air for at least 48 hours on your assignment. Above all, you must be a REGULAR on the repeater, a REGULAR when it comes to in-person training sessions, and a REGULAR in accruing additional training available to your unit.*

T2C10 What is the preamble in a formal traffic message?

A. The first paragraph of the message text.
B. The message number.
C. The priority handling indicator for the message.
D. The information needed to track the message as it passes through the amateur radio traffic handling system.

It is important to ***keep track of emergency messages*** as they pass through the well-structured amateur radio traffic-handling system. The make-up of the PREAMBLE of the message gives us the details to know where that message came from, and where it is going down the line. **ANSWER D.**

Hams are well-known for their work with the Red Cross, Salvation Army, and others providing emergency communications.

T2C08 What is usually considered to be the most important job of an amateur operator when handling emergency traffic messages?

A. Passing messages exactly as written, spoken or as received.
B. Estimating the number of people affected by the disaster.
C. Communicating messages to the news media for broadcast outside the disaster area.
D. Broadcasting emergency information to the general public.

A good emergency traffic handler will always write down their radio traffic in block letters, word for word. This way, when they ***pass the message*** on to authorities, it will be ***exactly as written, spoken as received*** - word for word! **ANSWER A.**

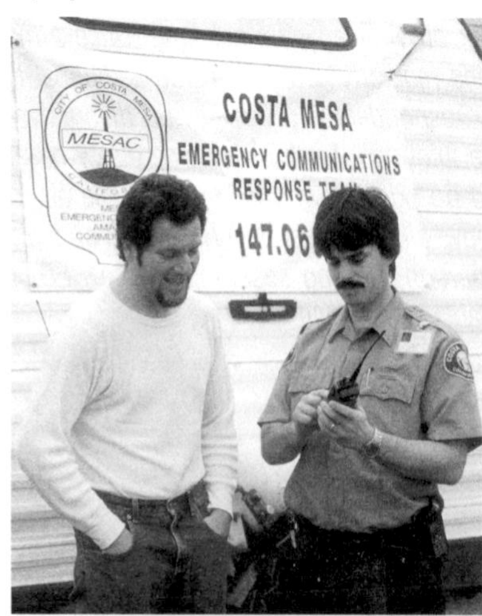

Hams Operating from an Emergency Communications Command Post.

T2C11 What is meant by the term "check" in reference to a formal traffic message?

A. The check is a count of the number of words or word equivalents in the text portion of the message.
B. The check is the value of a money order attached to the message.
C. The check is a list of stations that have relayed the message.
D. The check is a box on the message form that tells you the message was received.

Careful handling of format traffic requires ham radio message handlers to always include a "check" to make sure that all the ***words in a message*** indeed were received in their entirety.

ANSWER A.

When you're working emergency traffic, it's important to make sure every word and number is passed along just as you received it. A "check" helps assure accuracy.

▾ IF YOU'RE LOOKING FOR	▾ THEN VISIT
All About RACES	www.usraces.org
News on Emergency Groups	www.N4KSS.net/Reflectors.html
Military Radio Groups	www.netcom.Army.mil/MARS
More Military Groups	www.NAVYMARS.org
All About ARES	www.ARRL.org/ARES
ARRL Emergency Volunteers	www.ARRL.org/Volunteer

The Effect of the Ionosphere on Radio Waves

To help you with the questions on radio wave propagation, here is a brief explanation on the effect the ionosphere has on radio waves.

The ionosphere is the electrified atmosphere from 40 miles to 400 miles above the Earth. You can sometimes see it as "northern lights." It is charged up daily by the sun, and does some miraculous things to radio waves that strike it. Some radio waves are absorbed during daylight hours by the ionosphere's D layer. Others are bounced back to Earth. Yet others penetrate the ionosphere and never come back again. The wavelength of the radio waves determines whether the waves will be absorbed, refracted, or will penetrate and pass through into outer space. Here's a quick way to memorize what the different layers do during day and nighttime hours:

The D layer is about 40 miles up. The D layer is a Daylight layer; it almost disappears at night. D for Daylight. The D layer absorbs radio waves between 1 MHz to 7 MHz. These are long wavelengths. All others pass through.

The E layer is also a daylight layer, and it is very Eccentric. E for Eccentric. Patches of E layer ionization may cause some surprising reflections of signals on both high frequency as well as very-high frequency. The E layer height is usually 70 miles.

The F1 layer is one of the layers farthest away. The F layer gives us those Far away signals. F for Far away. The F1 layer is present during daylight hours, and is up around 150 miles. The F2 layer is also present during daylight hours, and it gives us the Furthest range. The F2 layer is 250 miles high, and it's the best for the Farthest range on medium and short waves. At nighttime, the F1 and F2 layers combine to become just the F layer at 180 miles. This F layer at nighttime will usually bend radio waves between 1 MHz and 15 MHz back to earth. At night, the D and E layers disappear.

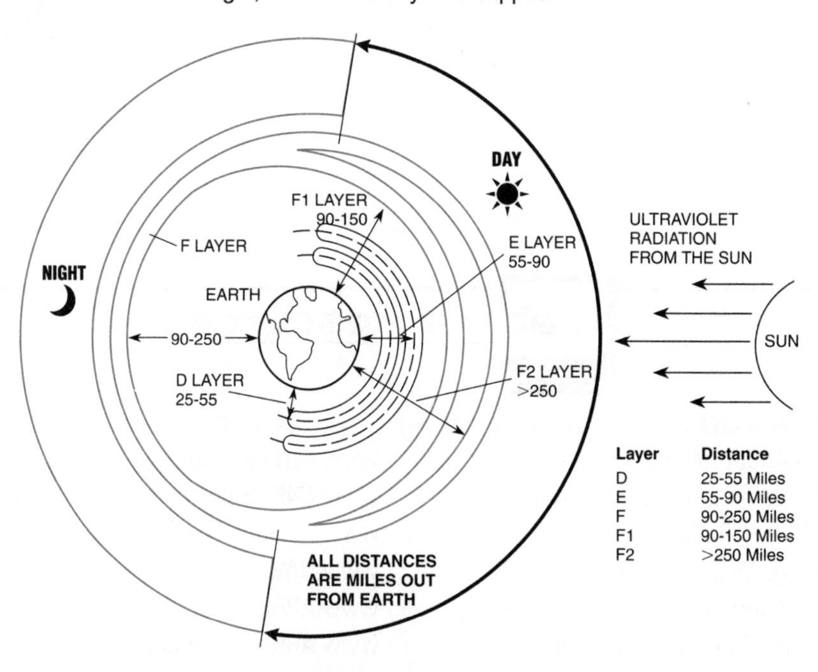

Layer	Distance
D	25-55 Miles
E	55-90 Miles
F	90-250 Miles
F1	90-150 Miles
F2	>250 Miles

Ionosphere Layers

Source: *Antennas — Selection and Installation,* © 1986, Master Publishing, Inc., Niles, Illinois

Weak Signal Propogation

T5C07 What is a usual name for electromagnetic waves that travel through space?

A. Gravity waves.
B. Sound waves.
C. Radio waves.
D. Pressure waves.

Electromagnetic waves are RADIO WAVES. A radio wave is composed of both an electric field and a magnetic field at right angles to each other. See the illustration on page 54. **ANSWER C.**

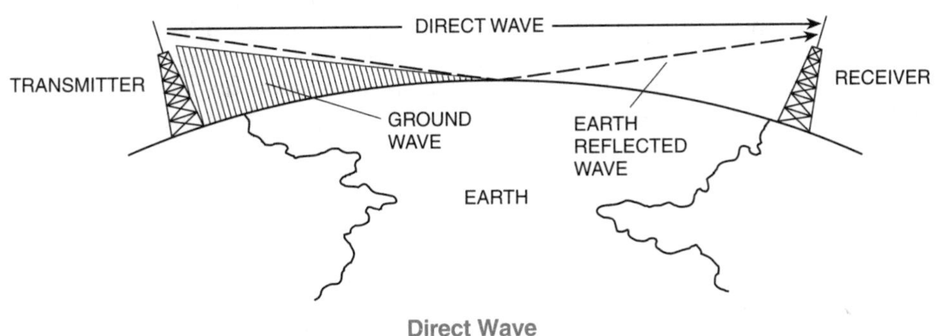

Direct Wave

Source: *Antennas — Selection and Installation,* © 1986, Master Publishing, Inc., Niles, Illinois

T3C10 What is the radio horizon?

A. The distance at which radio signals between two points are effectively blocked by the curvature of the Earth.
B. The distance from the ground to a horizontally mounted antenna.
C. The farthest point you can see when standing at the base of your antenna tower.
D. The shortest distance between two points on the Earth's surface.

On VHF and UHF frequencies, generally, your signals will travel "line of sight." Repeaters are usually put way up on buildings and mountain tops to extend line-of-sight range. But if you and a pal could walk on water, and you each started walking in the opposite direction, you would lose sight of each other's smiling face at about 5 miles due to the curvature of the Earth. Amazing, huh? ***VHF and UHF radio signals are blocked by the curvature of the Earth***, too. **ANSWER A.**

T3C11 Why do VHF and UHF radio signals usually travel somewhat farther than the visual line of sight distance between two stations?

A. Radio signals move somewhat faster than the speed of light.
B. Radio waves are not blocked by dust particles.
C. The Earth seems less curved to radio waves than to light.
D. Radio waves are blocked by dust particles.

Now if you and your pal, walking on water, lost sight of each other's smiling face at 5 miles because of the curvature of the Earth, you could probably keep yakking on your 2-meter handheld an additional couple of miles, thanks to the refractive index of air on radio signals, which makes the ***Earth seem less curved***. So if you do your line-of-sight calculations, add another 15 percent to your visual range and I bet your radio signals will be heard loud and clear beyond the visual horizon. **ANSWER C.**

T3C05 What is meant by the term "knife-edge" propagation?

A. Signals are reflected back toward the originating station at acute angles.
B. Signals are sliced into several discrete beams and arrive via different paths.
C. Signals are partially refracted around solid objects exhibiting sharp edges.
D. Signals propagated close to the band edge exhibiting a sharp cutoff.

You and a pal just earned your Technician Class licenses and you each have a new handheld, set to a simplex frequency. Your friend lives on the other side of a jagged rocky hill. Will you be able to communicate by radio, simplex (no repeater), over this hill? Likely, yes. VHF FM signals from base stations, handhelds, and mobile radios normally propagate vertical polarization. When the lower edge of the vertical wave strikes a distant, elevated ***sharp object***, like the cliff between you and your friend, the lower portion of the wave tends to drag over the cliff, ***causing the upper portion of the wave front to bend*** right down to where your friend is waiting to hear from you. Knife-edge propagation created by an intervening building or hill may still give you some great communications fun with you new license! **ANSWER C.**

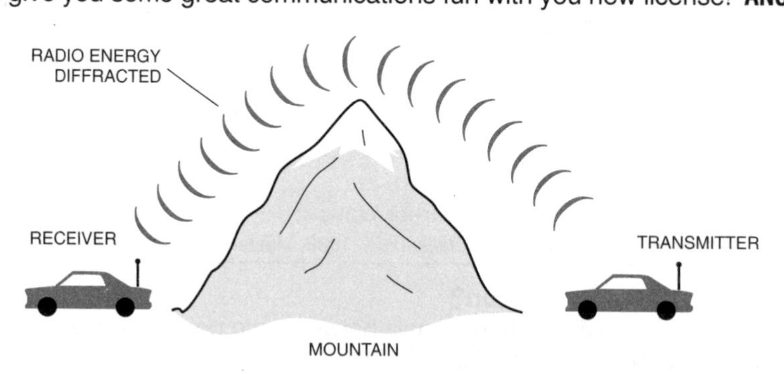

Knife-Edge Diffraction

T3C06 What mode is responsible for allowing over-the-horizon VHF and UHF communications to ranges of approximately 300 miles on a regular basis?

A. Tropospheric scatter.
B. D layer refraction.
C. F2 layer refraction.
D. Faraday rotation.

It is not uncommon to be able to communicate through a repeater several hundred miles away – well beyond line-of-sight. This is because your signal gets caught up in a warm air inversion, located in the troposphere. This longer than usual communications range is called ***tropospheric scatter***, or sometimes referred to as tropospheric ducting. **ANSWER A.**

T3C08 What causes "tropospheric ducting?"

A. Discharges of lightning during electrical storms.
B. Sunspots and solar flares.
C. Updrafts from hurricanes and tornadoes.
D. Temperature inversions in the atmosphere.

Ever see a mirage? This is frequently seen as water shimmering on a roadway, but it is actually the blue sky that you see on an inferior mirage. Another type of mirage may occur above us, called a superior mirage. This allows you to see things suspended upside down, many miles away! The superior mirage also creates longer than usual range on the 2 meter and 70 cm bands, thanks to a ***temperature inversion*** when a layer of warm air traps colder air below it creating a tropospheric duct. **ANSWER D.**

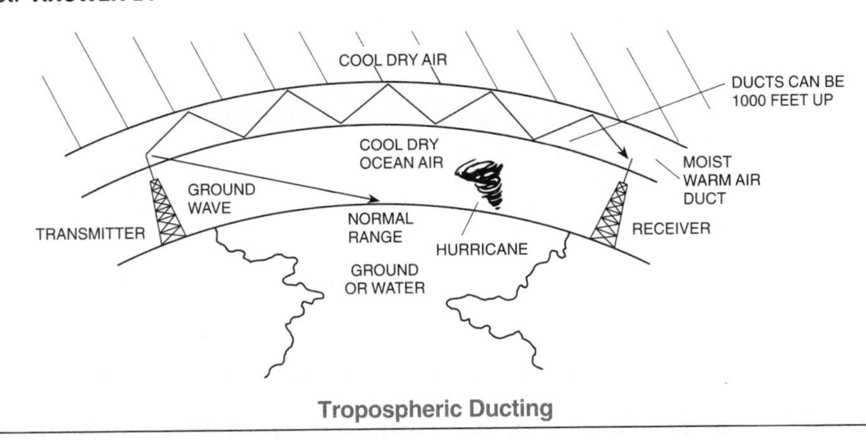

Tropospheric Ducting

T3C03 What is a characteristic of VHF signals received via auroral reflection?

A. Signals from distances of 10,000 or more miles are common.
B. The signals exhibit rapid fluctuations of strength and often sound distorted.
C. These types of signals occur only during winter nighttime hours.
D. These types of signals are generally strongest when your antenna is aimed to the south (for stations in the Northern Hemisphere).

VHF signals on the 2 meter band travel "line of sight." The higher up you are from the surface of the Earth, the better your range. An interesting phenomenon, called an auroral reflection, may influence 2 meter single sideband and FM signals. Six meter and 2 meter radio waves will bounce off an auroral curtain as if they were bounced off a steel building! The ***incoming signals*** from a distant station will ***sound fluttery and distorted***. This is the unmistakable sound of an auroral bounce that can be heard hundreds of miles away! Listen to an auroral radio call on the enclosed CD! **ANSWER B.**

T3C07 What band is best suited to communicating via meteor scatter?

A. 10 meters.
B. 6 meters.
C. 2 meters.
D. 70 cm.

The Leonids and Geminids meteor showers are fun for hams! As a new Technician Class operator, you can work the entire ***6 meter band***, running up to 1500 watts of power output. This is plenty of power to bounce a signal off a meteor trail or even the moon! Hear this on the enclosed CD. **ANSWER B.**

T3A11 Which part of the atmosphere enables the propagation of radio signals around the world?

A. The stratosphere.
B. The troposphere.
C. The ionosphere.
D. The magnetosphere.

It is the ionosphere that gives us propagation all over North America on the 6- and 10-meter bands, and as we climb higher on Solar Cycle 24, we may even get some 6- and 10-meter propagation halfway around the world! Thanks to the ***ionosphere***! **ANSWER C.**

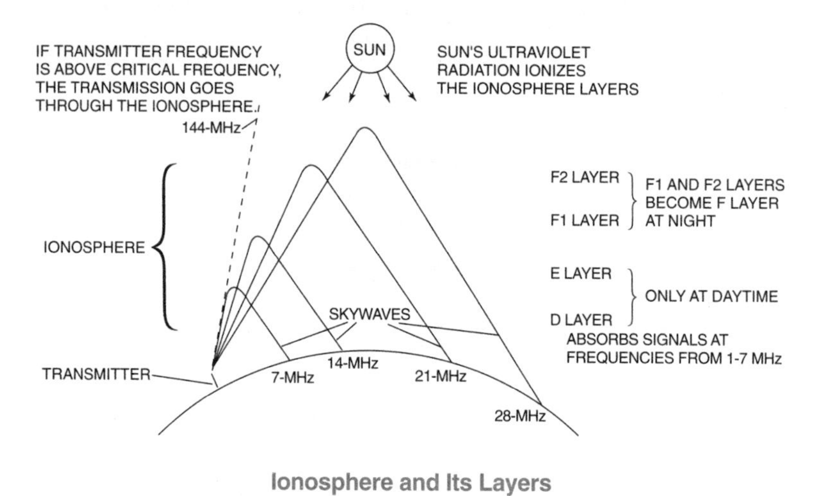

Ionosphere and Its Layers

T3C02 Which of the following might be happening when VHF signals are being received from long distances?

A. Signals are being reflected from outer space.
B. Signals are arriving by sub-surface ducting.
C. Signals are being reflected by lightning storms in your area.
D. Signals are being refracted from a sporadic E layer.

The 6-meter band, 50 MHz to 54 MHz, is referred to as THE MAGIC BAND. During the summertime, it is full of surprises. Many Technician Class operators will buy 6-meter multi-mode equipment and operate upper sideband at 50.125 MHz. So, one morning you are on the air chatting with a pal across town, and a voice out of nowhere drops into the conversation and announces that she is located 1,500 miles away. Her signal is coming in via ***Sporadic-E refractions*** off of ionized patches of the ionospheric E-layer. This is a common occurrence on the 6- and 10-meter bands, during summertime, in the early morning and late afternoon hours. **ANSWER D.**

T3C09 What is generally the best time for long-distance 10 meter band propagation?

A. During daylight hours.
B. During nighttime hours.
C. When there are coronal mass ejections.
D. Whenever the solar flux is low.

Remember your ***10 meter*** SSB privileges are from 28.300 to 28.500 MHz. Start out around 28.400 and give your best spirited CQ call. Likely, ***during the day***, you may get a response from another station halfway across the country! On 6 meters, try your SSB CQ call on 50.125 MHz, upper sideband. **ANSWER A.**

Ham Hint: *Did you spot that audio CD in the front of this book? Go ahead and play it! This CD has plenty of sounds of different sky wave skip communications. This CD plays in a regular audio CD system and will give you lots of the sounds of excitement of ham radio Technician Class operating. Play it now!*

T3A09 Which of the following is a common effect of "skip" reflections between the Earth and the ionosphere?

A. The sidebands become reversed at each reflection.
B. The polarization of the original signal is randomized.
C. The apparent frequency of the received signal is shifted by a random amount.
D. Signals at frequencies above 30 MHz become stronger with each reflection.

As a new Technician Class operator, you probably won't hear any ionospheric reflections, called "skip," on 2 meters or 70 cm. But down on 6 meters, with a multimode mobile or base station, there is plenty of summertime skip. Skip happens when the signals refract and reflect off the ionosphere, and distant stations up to 1,000 miles away come booming in. About every 30 seconds, that distant station will go from strong to very weak and then back to strong. This is caused by the ***random, ever changing polarization of the original signal***. **ANSWER B.**

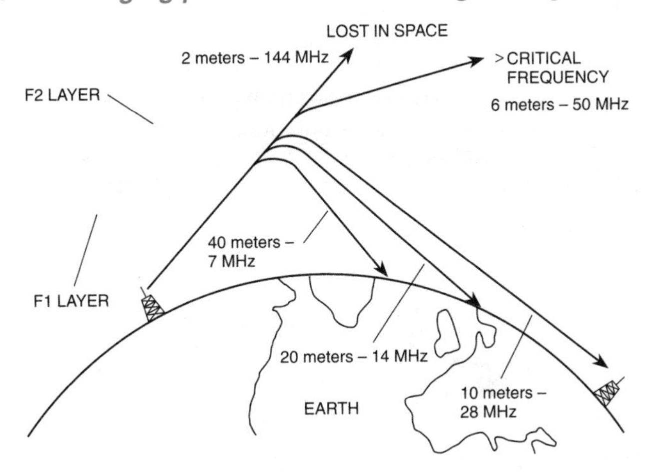

Critical Frequency

Source: *Antennas — Selection and Installation,* © 1986, Master Publishing, Inc.

T3C04 Which of the following propagation types is most commonly associated with occasional strong over-the-horizon signals on the 10, 6, and 2 meter bands?

A. Backscatter.
B. Sporadic E .
C. D layer absorption.
D. Gray-line propagation.

Regularly on 10 meters we may communicate with sky wave stations close to 1,000 miles away, thanks to ***sporadic "E skip."*** On 6 meters, this phenomenon usually occurs several times a week during the summertime and in December. On 2 meters, skip conditions may prevail once or twice a month, just for a few minutes, during the summertime. **ANSWER B.**

T3C01 Why are "direct" (not via a repeater) UHF signals rarely heard from stations outside your local coverage area?

A. They are too weak to go very far.
B. FCC regulations prohibit them from going more than 50 miles.
C. UHF signals are usually not reflected by the ionosphere.
D. They collide with trees and shrubbery and fade out.

The 6- and 10-meter bands are exceptions to Technician Class privileges that usually lead to signals that travel line of sight. On the 6- and 10-meter bands, during the summer months, signals will many times refract off the E-layer of the ionosphere, and come back down hundreds and thousands of miles away. However, on 2 meters and higher on ***UHF*** (ultra-high frequency), ***signals are usually not reflected*** nor refracted by the ionosphere. **ANSWER C.**

T3A08 What is the cause of irregular fading of signals from distant stations during times of generally good reception?

A. Absorption of signals by the "D" layer of the ionosphere.
B. Absorption of signals by the "E" layer of the ionosphere.
C. Random combining of signals arriving via different path lengths.
D. Intermodulation distortion in the local receiver.

You are on the air with your new dual-band handheld radio. Try to keep the antenna as vertical as possible, and move around until you hear the distant repeater coming in strong. This is called a "hot spot" where you have the best reception. Move just a little bit one way or another, and reception may drop dramatically, which we will call a "NOT spot". Something interesting may happen if you live near enough to an airport; your hot spot to that distant repeater will occasionally encounter brief signal fades. This is called multipath, and can be created by airplanes, and sometimes passing cars and busses, which give the ***distant incoming signal different paths to your handheld***. If the direct path and the alternate paths are out of phase, fading may occur. **ANSWER C.**

IF YOU'RE LOOKING FOR	THEN VISIT
Info on Microwave Groups	www.MicrowaveUpdate.org
More on Microwave	www.Ham-Radio.com
6 Meter Info	www.smirk.org/
See Aurora upclose	www.spaceweather.com
Tropo ducting reporting site	http://dx.qsl.net/propagation/tropo.php
WWV propagation beacon information	http://tf.nist.gov/stations/wwv.html
Educational site about propagation	http://prop.hfradio.org
Latest propagation conditions info	http://wap.hfradio.org
More on propagation	www.haarp.alaska.edu

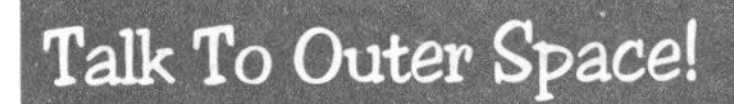

Talk To Outer Space!

T1A05 What is the FCC Part 97 definition of a space station?

A. Any multi-stage satellite.
B. An Earth satellite that carries one of more amateur operators.
C. An amateur station located less than 25 km above the Earth's surface.
D. An amateur station located more than 50 km above the Earth's surface.

Special rules pertain to amateur operation in a space station. ***A space station is considered any amateur station located more than 50 kilometers above the Earth's surface***. Remote control of a model aircraft at 1000 feet is not considered a space station. [97.3(a)(40)] **ANSWER D.**

T8B04 Which amateur stations may make contact with an amateur station on the International Space Station using 2 meter and 70 cm band amateur radio frequencies?

A. Only members of amateur radio clubs at NASA facilities.
B. Any amateur holding a Technician or higher class license.
C. Only the astronaut's family members who are hams.
D. You cannot talk to the ISS on amateur radio frequencies.

What a thrill to talk on your small handheld to an astronaut! It happens a lot, and with the ***Technician Class license or higher*** you, too, can explore space with astronauts and cosmonauts. **ANSWER B.**

Ham Hint: *If you're wondering how you can easily tune into the International Space Station, that is easy with a little 2-meter FM handheld. The International Space Station downlink, FM, is 145.800. When it is passing over, you might hear an astronaut and you might hear packet stations relaying messages globally. That's right, your little handheld is plenty powerful enough to pick up the International Space Station on an overhead pass!*

The International Space Station has a big ham station on board.
Photo courtesy of N.A.S.A.

T8B03 Which of the following can be done using an amateur radio satellite?

A. Talk to amateur radio operators in other countries.
B. Get global positioning information.
C. Make telephone calls.
D. All of these choices are correct.

Satellite signals know no political boundaries, and it is perfectly okay to ***speak with ham satellite users in other countries***. Look at your satellite-tracking software and see the illuminated "footprint" of where you can expect your signal to end up over 1,000 miles away! **ANSWER A.**

T8B10 What do the initials LEO tell you about an amateur satellite?

A. The satellite battery is in Low Energy Operation mode.
B. The satellite is performing a Lunar Ejection Orbit maneuver.
C. The satellite is in a Low Earth Orbit.
D. The satellite uses Light Emitting Optics.

Ham satellites are not geostationary, but rather in ***Low Earth Orbits***, which reduces the amount of power we need to hear our own signals coming back. The low earth orbit also diminishes the delay echo and, depending on the satellite orbit, you could have a conversation for just a few minutes, or sometimes a few hours. Some orbits are elliptical, and as long as you and the other station can both still see the satellite, you are on the air! **ANSWER C.**

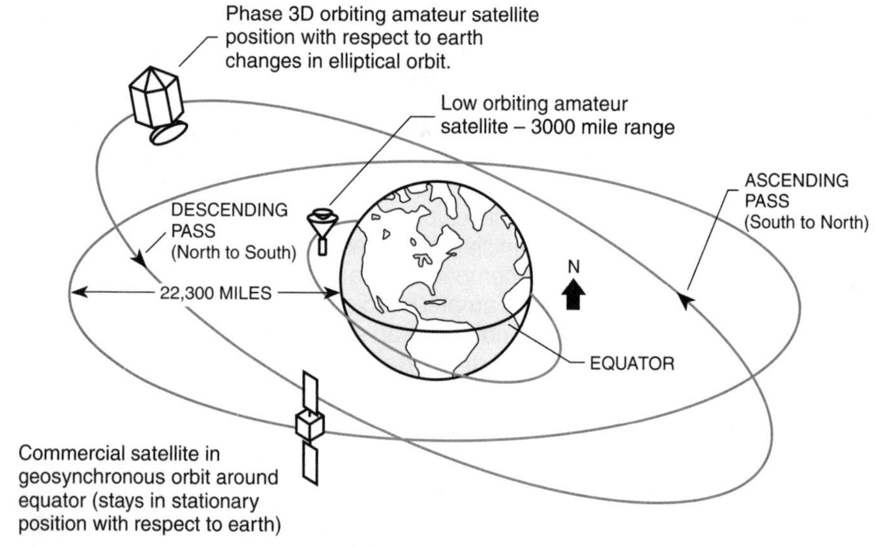

Orbiting Satellites

Ham Hint: *AMSAT has plans to deploy a satellite called Eagle, which will be in a high earth orbit (HEO) for a wideband transponder with a footprint that can cover a significant portion of the Earth's surface. AMSAT also is working on "Express" that will be supporting amateur radio operators with a constellation of satellites that will open new communication opportunities. But for all this to happen, AMSAT deserves all of our support. We hope all of you will support AMSAT, the nonprofit ham radio organization that sponsors our satellite programs. www.AMSAT.org*

T8B06 What can be used to determine the time period during which an amateur satellite or space station can be accessed?

A. A GPS receiver.
B. A field strength meter.
C. A telescope.
D. A satellite tracking program.

The Radio Amateur Satellite Corporation offers ***satellite-tracking programs***, plus programs from other providers, to show all ham radio satellite positions. You also will see the International Space Station as one of the heavenly bodies capable of relaying ham radio signals. I encourage all of you to log onto the AMSAT web page and join up to help promote more satellites in space. **ANSWER D.**

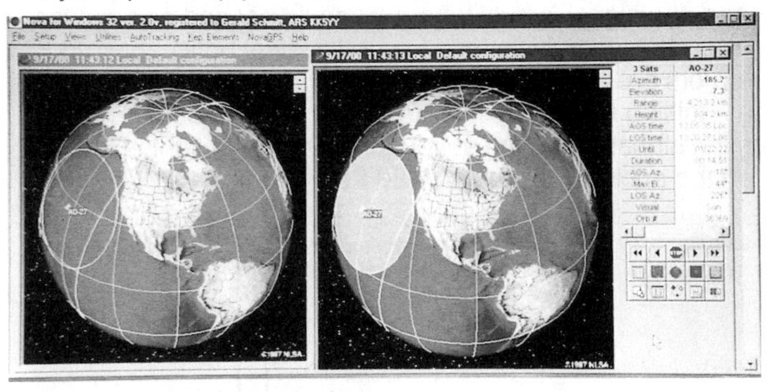

Computer programs and websites can show you where and when an amateur satellite or the Space Station will be in range of your ham station.

T8B05 What is a satellite beacon?

A. The primary transmit antenna on the satellite.
B. An indicator light that that shows where to point your antenna.
C. A reflective surface on the satellite.
D. A transmission from a space station that contains information about a satellite.

The satellite beacon is a continuous faint ***signal that carries digitized information about the satellite*** itself in orbit. Ham satellite controllers can monitor the beacon with special software and check everything from the inside temperature to how fully-charged the satellite batteries are. **ANSWER D.**

T8B09 What causes "spin fading" when referring to satellite signals?

A. Circular polarized noise interference radiated from the sun.
B. Rotation of the satellite and its antennas.
C. Doppler shift of the received signal.
D. Interfering signals within the satellite uplink band.

As you are listening to an FM or SSB satellite downlink signal, it will rapidly rise and fall in signal strength rapidly. This is called ***"spin fading"***. It is because the ***satellite is rotating*** in space, in order to keep its solar panels from overheating. This rotation in space makes the signals fade in and out. **ANSWER B.**

Tracking and communicating through amateur satellites can be done with a cross-polarized satellite antenna

T8B07 With regard to satellite communications, what is Doppler shift?

A. A change in the satellite orbit.
B. A mode where the satellite receives signals on one band and transmits on another.
C. An observed change in signal frequency caused by relative motion between the satellite and the Earth station.
D. A special digital communications mode for some satellites.

Ham satellite operators are constantly turning the dial to make up for Doppler shift. If the satellite is headed towards you, the received signal may appear a few kilohertz high. As it passes overhead, the signal is right where it should be; and as it moves away from you, the signal will be several kilohertz lower. It is very noticeable, and experienced satellite operators pre-store memory channels that will compensate for ***Doppler shift***. Amazing – talking through a satellite where you can actually hear ***frequency changes*** as the signal is coming at you at the speed of light! Hear a sample of Doppler shift on the enclosed CD! **ANSWER C.**

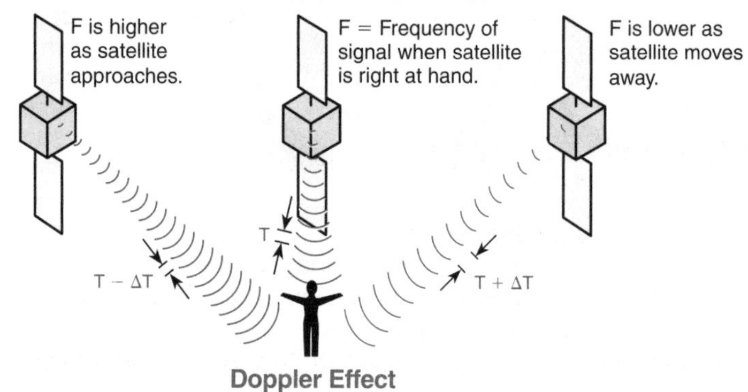

Doppler Effect

T8B08 What is meant by the statement that a satellite is operating in "mode U/V"?

A. The satellite uplink is in the 15 meter band and the downlink is in the 10 meter band.
B. The satellite uplink is in the 70 cm band and the downlink is in the 2 meter band.
C. The satellite operates using ultraviolet frequencies.
D. The satellite frequencies are usually variable.

An exciting Technician Class privilege is working ham radio satellites, called OSCARs. You will need a full-duplex, dual-band radio to give you the best access to those OSCARs that transmit and receive FM signals. If the mode is "U/V" it means we transmit on the ***satellite uplink frequency on the 70 cm "U" band***, and we listen to the satellite ***downlink on the 2 meter "V" band.*** **ANSWER B.**

Frequency Bands	Frequency Range	Modes
High Frequency	21 - 30 MHz	Mode H
VHF	144 - 146 MHz	Mode V
UHF	435 - 438 MHz	Mode U
L band	1.26 - 1.27 GHz	Mode L
S band	2.4 - 2.45 GHz	Mode S
C band	5.8 GHz	Mode C
X band	10.4 GHz	Mode X
K band	24 GHz	Mode K

T8B02 How much transmitter power should be used on the uplink frequency of an amateur satellite or space station?

A. The maximum power of your transmitter.
B. The minimum amount of power needed to complete the contact.
C. No more than half the rating of your linear amplifier.
D. Never more than 1 watt.

Each of the approximate 15 satellites in orbit may carry different uplink and downlink modes. It's always best to run the ***minimum amount of power*** through the satellite so as not to overload the satellite electronics onboard with a too-strong signal. If you are operating voice, and tune in your own voice echo, start pulling your power back until you noticeably fade away from the downlink. Then increase power slightly, and give a CQ and see who is hearing your transmission over 1,000-plus miles away! [97.313(a)] **ANSWER B.**

To work satellites with your handheld, buy a small directional antenna for your satellite radio. You probably won't hear much with your rubber duck antenna.

T1A07 What is the FCC Part 97 definition of telemetry?

A. An information bulletin issued by the FCC.
B. A one-way transmission to initiate, modify or terminate functions of a device at a distance.
C. A one-way transmission of measurements at a distance from the measuring instrument.
D. An information bulletin from a VEC.

Good news, our orbiting ham satellites (called OSCARs) are having a good day in space!

Battery condition: Full
Outside temperature: Very cold
Power output: Excellent
Solar Panels: Bring on the sun!

The way we determine the operating parameters of a distant ham satellite, or a mountaintop ham radio repeater and weather station, is through a one way transmission beaming down ***telemetry*** from its internal ***measuring*** devices. [97.3(a)(45)] **ANSWER C.**

T1A06 What is the FCC Part 97 definition of telecommand?

A. An instruction bulletin issued by the FCC.
B. A one-way radio transmission of measurements at a distance from the measuring instrument.
C. A one-way transmission to initiate, modify or terminate functions of a device at a distance.
D. An instruction from a VEC.

The word ***"telecommand"*** refers to the ***one way control transmission to turn on, turn off, or modify*** options of another ham radio device many miles away – sometimes thousands of miles away!

Turning ON an amateur radio satellite
Initiating a satellite mode change
Turning OFF a distant propagation radio beacon
Changing data ports on a distant mountaintop digital repeater system

You get the idea – sending a telecommand signal to that mountaintop repeater tied into the internet, changing ham radio internet nodes. [97.3(a)(43)] **ANSWER C.**

Website Resources

▼ IF YOU'RE LOOKING FOR	▼ THEN VISIT
Join AMSAT	www.AMSAT.org
Digital Groups	www.TAPR.org
IRLP Info	status.irlp.net
All about IRLP	www.IRLP.net
Echolink info	www.echolink.org
Winlink info	www.winlink.org
Yaesu's WIRES	www.VXSTD.com
GPS info	www.geosat.us www.avmapnavigation.com www.avmap.us
PSK 31 info DATA Readers	www.digipan.net
LINUX info	RADIO.LINUX.org.AU
Find Next Satellite Pass	www.Heavens-Above.com
Cloning software	www.cloningsoftware.com
QRP and hombrew links	www.AL7FS.us
Handheld Satellite Antenna	www.arrowantennas.com
Working satellites with a hand-held radio	www.work-sat.com
Satellite software tracking applications	www.bigfattail.com
Satellite tracking Software for Macs!	www.dogparksoftware.com

CD 3 **TRACK 3**

Your Computer Goes Ham Digital!

T8D09 What code is used when sending CW in the amateur bands?

A. Baudot.
B. Hamming.
C. International Morse.
D. Gray.

We use good old fashioned ***International Morse code*** when sending ***CW*** (continuous wave signals) on the ham bands. You may use a straight key or an automatic keyer. I like the straight key because it helps me form good dits and dahs as I'm learning to send Morse code, CW. **ANSWER C.**

T8D10 Which of the following can be used to transmit CW in the amateur bands?

A. Straight Key.
B. Electronic Keyer.
C. Computer Keyboard.
D. All of these choices are correct.

If you decide to do some CW sending, GREAT! You can use a ***straight key***, or an ***electronic keyer***. You could even generate CW with a ***computer*** – but don't expect that computer to be able to read CW letters as well as your good old ears and brain. **ANSWER D.**

T8D01 Which of the following is an example of a digital communications method?

A. Packet.
B. PSK31.
C. MFSK.
D. All of these choices are correct.

Some new Technician Class operators start out, on the air, exclusively with their computers and a small ham radio transceiver. A handheld could allow Packet communications, similar to short message texting, but on ham channels. A larger, multi-mode base station tied in to your computer's sound card could lead to exciting PSK31 worldwide data communications on the 10 meter band. And for those of you who enjoy bouncing signals off the mood and meteors, MFSK is a fabulous digital VHF/UHF mode where your computer does all the work of sorting signals out of the noise! ***MFSK, PACKET, and PSK31 data modes*** let your laptop, palm, and home computer do all the signaling, digitally! **ANSWER D.**

T4A06 Which of the following would be connected between a transceiver and computer in a packet radio station?

A. Transmatch.
B. Mixer.
C. Terminal node controller.
D. Antenna.

In between your handheld or mobile radio and the computer is a device called a ***terminal node controller***, abbreviated TNC. Some mobile and handheld radios already have the TNC built in! Your little handheld now plugs into the terminal node controller, and the TNC then hooks up to your tiny PDA or laptop for text messaging. Packet radio can be used for many important purposes in ham radio – an emergency responder bulletin board; packet cluster reports on noteworthy, long-range stations showing up on the radio dial; PDA-to-PDA wireless messaging; and my favorite, automatic position reporting system. Packet bursts send out my latitude and longitude to other stations, some of which are digipeaters that relay my position on the Internet for everyone to see where I am. Additional features of APRS include remote weather station monitoring, short text messaging, emergency alert, and the tie-in to a vehicle navigation device to graphically see where other APRS users are located around you on the map! **ANSWER C.**

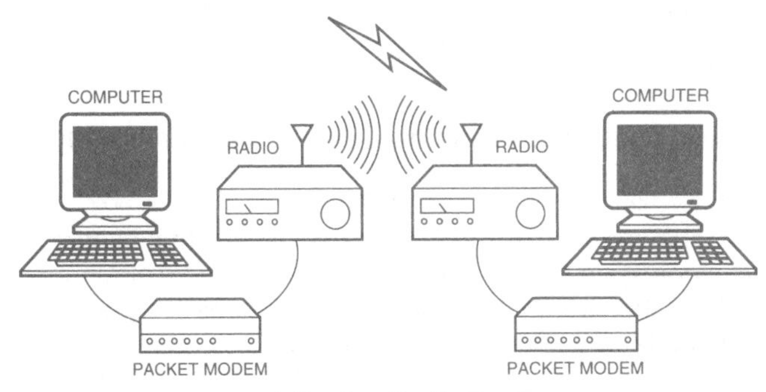

A Packet Radio System

T4A07 How is the computer's sound card used when conducting digital communications using a computer?

A. The sound card communicates between the computer CPU and the video display.
B. The sound card records the audio frequency for video display.
C. The sound card provides audio to the microphone input and converts received audio to digital form.
D. All of these choices are correct.

Ham radios and computers are made for each other! If your computer has a built in sound card, you can "see" what is sent via digital modes like CW, PSK-31, radioteleprinter, and nearly a dozen other digital modes, most of which sound like "hash" to your ear when listening to your receiver. Take the ***sounds*** of this "hash", feed them ***into the mic input jack*** on your computer's sound card, run the appropriate ***decoder*** software, and watch ***digital messages*** filling in on your screen in text, and maybe even graphics ! Now, type in your outgoing message and your sound card will generate the digital "hash," which is fed into a signal level matching box that connects to your radio's mic input, and also keys your radio on to transmit. Visit www.westmountainradio.com to visit to see the multi-rig interface equipment that goes between your ham gear and computer . Let you laptop sound card with appropriate software do the decoding of digital messages! **ANSWER C.**

T8D08 Which of the following may be included in packet transmissions?

A. A check sum which permits error detection.
B. A header which contains the call sign of the station to which the information is being sent.
C. Automatic repeat request in case of error.
D. All of these choices are correct .

Get your computers ready for some Technician Class fun. Packet radio is a great way to send blocks of information, quickly, over the airwaves. Your packet will include a header which contains the call sign of the station you wish to reach, and several steps of error detection and error repeat requests. So, for this question, ***all of these features*** are included when you send packet radio. **ANSWER D.**

T8D11 What is a "parity" bit?

A. A control code required for automatic position reporting.
B. A timing bit used to ensure equal sharing of a frequency.
C. An extra code element used to detect errors in received data.
D. A "triple width" bit used to signal the end of a character.

The parity bit is ***an extra data element used to detect errors*** in received data. It's good in making sure the copy you are reading is indeed what was sent. **ANSWER C.**

T7B12 What does the acronym "BER" mean when applied to digital communications systems?

A. Baud Enhancement Recovery.
B. Baud Error Removal.
C. Bit Error Rate.
D. Bit Exponent Resource.

If you are into computers, you are going to love your new Technician Class license privileges using data. The term BER stands for ***Bit Error Rate***. Hopefully, your BER will be very low! That means your throughput will be high! **ANSWER C.**

T3A10 What may occur if VHF or UHF data signals propagate over multiple paths?

A. Transmission rates can be increased by a factor equal to the number of separate paths observed.
B. Transmission rates must be decreased by a factor equal to the number of separate paths observed.
C. No significant changes will occur if the signals are transmitting using FM.
D. Error rates are likely to increase.

You are going to have a blast sending data with your laptop and your ham set. When conditions are good, data exchanges are letter/number perfect! However, when ***multipath signals*** add and subtract to your signal, ***error rates are likely to increase***. **ANSWER D.**

T8B11 What is a commonly used method of sending signals to and from a digital satellite?

A. USB AFSK.
B. PSK31.
C. FM Packet.
D. WSJT.

We can operate digital communications through satellites and through the International Space Station. ***FM packet*** is one of the most popular digital communication systems. It is fun to receive FM packet transmissions from other hams, usually stored and forwarded, via the satellite or space station. **ANSWER C.**

T8D02 What does the term APRS mean?

A. Automatic Position Reporting System.
B. Associated Public Radio Station.
C. Auto Planning Radio Set-up.
D. Advanced Polar Radio System.

Now that GPS equipment is sized down to fit into a cell phone, the ham radio with a built-in TNC for APRS allows you to ***automatically send your latitude and longitude*** every couple of minutes with nothing more than a little VHF handheld and a postage-stamp-sized GPS receiver. **ANSWER A.**

Kenwood dual bander plugged into the Avmap G5 GPS position plotter.

T8D03 Which of the following is normally used when sending automatic location reports via amateur radio?

A. A connection to the vehicle speedometer.
B. A WWV receiver.
C. A connection to a broadcast FM sub-carrier receiver.
D. A Global Positioning System receiver.

The output of your ***GPS receiver*** is an NMEA 0183 format that will connect to the terminal node controller either built-in or external to your handheld or mobile radio. **ANSWER D.**

T8D06 What does the abbreviation PSK mean?

A. Pulse Shift Keying.
B. Phase Shift Keying.
C. Packet Short Keying.
D. Phased Slide Keying.

PSK stands for ***Phased Shift Keying***; and if you listen carefully, the little warble is indeed information passing over an ultra-narrow carrier. Hear the sounds of PSK on the enclosed CD! **ANSWER B.**

T8D07 What is PSK31?

A. A high-rate data transmission mode.
B. A method of reducing noise interference to FM signals.
C. A method of compressing digital television signal.
D. A low-rate data transmission mode .

Down on the 10-meter band, a multi-mode radio may hear the sound of a constant whistle. If you listen carefully, you can even hear a slight warble, and that warble is conveying information on a continuous carrier called PSK 31. ***PSK 31*** is a remarkable digital mode that slices through interference and gets the message across – sometimes from here to the moon and back again! While the ***transmission rate is about normal typing speed***, it is a slow but adequate speed to converse with other hams via your computer and multi-mode radio on many of the Technician Class authorized frequencies. **ANSWER D.**

You can connect a PSK-31 and RTTY data reader to your radio to decode messages.

T8C11 What name is given to an amateur radio station that is used to connect other amateur stations to the Internet?

A. A gateway.
B. A repeater.
C. A digipeater.
D. A beacon.

Hundreds of ham radio operators throughout the world have tied their home computers and ham sets into the Internet. The system is called WinLink 2000, EchoLink, and IRLP. These free ***gateway*** stations are fully automated to provide ham radio operators access to the Internet for sending and receiving e-mails. **ANSWER A.**

Ham Hint: *For the Technician Class operator, VHF and UHF gateway stations could allow the ham radio emergency communicator, in the field, to send messages via the Internet to an American Red Cross center or an EOC. This system also allows non-licensed hams to generate e-mail to you, going through a WinLink gateway station. And once you get your General Class license, you and I could cruise the world and always have e-mail capabilities over longer-range, high-frequency networks. While bandwidth limitations do not support surfing the web, you can download and send e-mails along with VHF/UHF medium-resolution diagrams and charts, too.*

T8C09 How might you obtain a list of active nodes that use VoIP?

A. From the FCC Rulebook.
B. From your local emergency coordinator.
C. From a repeater directory.
D. From the local repeater frequency coordinator.

The Internet is your best source of for a voice-over-Internet node directory. Also, your local repeater control operator offering VoIP may have some favorite nodes to suggest, too. ***Repeater directories*** also list VoIP nodes. **ANSWER C.**

T8C10 How do you select a specific IRLP node when using a portable transceiver?

A. Choose a specific CTCSS tone.
B. Choose the correct DSC tone.
C. Access the repeater autopatch.
D. Use the keypad to transmit the IRLP node ID.

Most new handhelds have a minimum of 9 programmable memories for speed-dial number sending. ***Store your favorite IRLP nodes*** in your radio's speed-dial and you won't need to remember all of the node numbers and manually tap them in ***on the keypad***. **ANSWER D.**

T8C08 What is required in place of on-air station identification when sending signals to a radio control model using amateur frequencies?

A. Voice identification must be transmitted every 10 minutes.
B. Morse code ID must be sent once per hour.
C. A label indicating the licensee's name, call sign and address must be affixed to the transmitter.
D. A flag must be affixed to the transmitter antenna with the station call sign in 1 inch high letters or larger.

If you have a multi-mode, 6-meter radio, tune around 50.800 to 51.0 MHz, and 53.1 to 53.9 MHz. Chances are you'll hear some strange radio control signals, and they don't need to be identified in Morse code because the ***only required identification is a label*** indicating your ham call sign and address that is affixed on the actual 1-watt transmitter. [97.215(a)] **ANSWER C.**

T8C07 What is the maximum power allowed when transmitting telecommand signals to radio controlled models?

A. 500 milliwatts.
B. 1 watt.
C. 25 watts.
D. 1500 watts.

Ham radio operators are permitted specific channels on our own 6-meter band to control radio controlled models. This is a blessing – no longer will your airplane take a nosedive when somebody accidentally hits their garage door opener! The ***maximum power*** allowed for controlling your model aircraft on the ham radio 6-meter band is ***1 watt***. One watt will allow you to fly your model to a point you won't even be able to see it in the sky! Better have your name and address on it! [97.215(c)] **ANSWER B.**

Hams can use frequencies on the 6-Meter Band to radio control a model aircraft.

T8D04 What type of transmission is indicated by the term NTSC?

A. A Normal Transmission mode in Static Circuit.
B. A special mode for earth satellite uplink.
C. An analog fast scan color TV signal.
D. A frame compression scheme for TV signals.

Here in the US, ***NTSC is a standard for fast-scan color television signals*** for analog ham radio ATV stations. TV broadcasters have made the switch to all-digital, so hams are the last ones to stick with the tried-and-proven NTSC analog transmissions. So don't throw out your old analog TV set just yet – with a simple $99 down-converter, fast-scan ham television in full color awaits you! (FYI, NTSC stands for National Television System Committee.) **ANSWER C.**

When you're ready, you can add the fun of ATV to your ham shack.

Multi-Mode Radio Excitement

T7A09 Which of the following devices is most useful for VHF weak-signal communication?

A. A quarter-wave vertical antenna.
B. A multi-mode VHF transceiver.
C. An omni-directional antenna.
D. A mobile VHF FM transceiver.

Your Technician Class license puts you in the mainstream of exciting, 2-meter and 432-MHz operation. You can bounce signals off of the Moon with your Technician Class license, and talk with stations hundreds of miles away off of meteor trails. Perhaps work the amateur satellites, or try sending signals thousands of miles within atmospheric temperature inversions. This excitement is available to you as a Technician operator, with just a simple FM transceiver. But it takes a ***multi-mode radio*** with CW and SSB capabilities to work weak signal VHF activities. Look for a multi-mode VHF transceiver to get started on the 2-meter band, using CW and upper sideband for weak signal work. There is also a satellite weak signal "window" between 145.800 to 146.000 MHz. **ANSWER B.**

What's a multi-mode transceiver?

It's a transceiver that can send and receive different modes of radio signals. The term "mode" refers to the type of modulation emission used to send signals. Look at the illustration below. We send Morse code signals using CW, continuous wave. When you tap out a message on a code key, you interrupt (turn off and on) the unmodulated carrier continuous wave. In AM modulation, the RF carrier wave is modulated by the information signal and the amplitude (height) of the signal varies. SSB is single sideband, a form of AM, amplitude modulation. FM is frequency modulation, and that's what most VHF/UHF hand-held radios use. In FM, the frequency of the carrier wave is changed by the information signal. To learn more about how radios work, I suggest Basic Communications Electronics *by Jack Hudson, W9MU and Jerry Luecke, KB5TZY, available from The W5YI Group, 800-669-9594 or www.w5yi.org.*

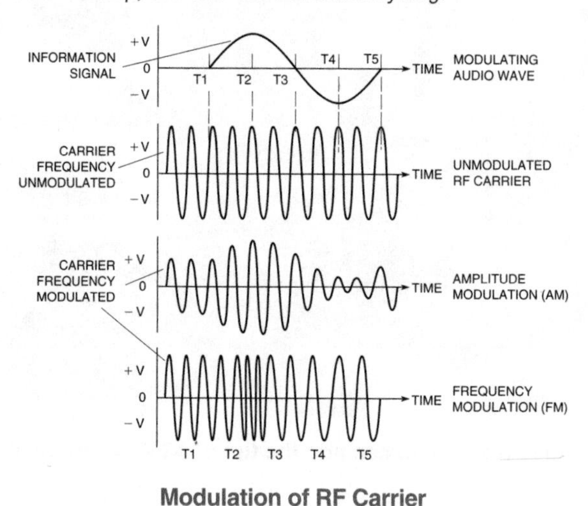

Modulation of RF Carrier

T8A05 Which of the following types of emission has the narrowest bandwidth?

A. FM voice.
B. SSB voice.
C. CW.
D. Slow-scan TV.

The Federal Communications Commission may soon rule on the switch from restricted types of communications over to the more reasonable bandwidth limitations for amateur communications. FM voice and slow-scan TV are modest bandwidth, SSB voice has a "skinny" bandwidth, and the ultra-narrowest bandwidth of all (of these answers) is ***Morse code, CW***. Even though the code test for CW has been eliminated for all classes of ham licenses, CW will always be a popular, ultra-narrow-bandwidth way of communicating. **ANSWER C.**

T8A11 What is the approximate maximum bandwidth required to transmit a CW signal?

A. 2.4 kHz.
B. 150 Hz.
C. 1000 Hz.
D. 15 kHz.

I hope all of you will try some CW down on the worldwide portion of your Technician Class privileges on 80-, 40-, 15- and 10-meters. When you switch your multi-mode rig over to CW, you will see that the bandwidth narrows down. This minimizes the pickup of other off frequency ***CW signals***. ***150 Hz*** is just right. **ANSWER B.**

T7A05 What is the function of block 1 if figure T4 is a simple CW transmitter?

A. Reactance modulator.
B. Product detector.
C. Low-pass filter.
D. Oscillator.

The oscillator stage generates the radio signal at a specific frequency. The oscillator has a built-in feedback circuit that keeps the ***oscillator*** stage in amplification of the radio signal. **ANSWER D.**

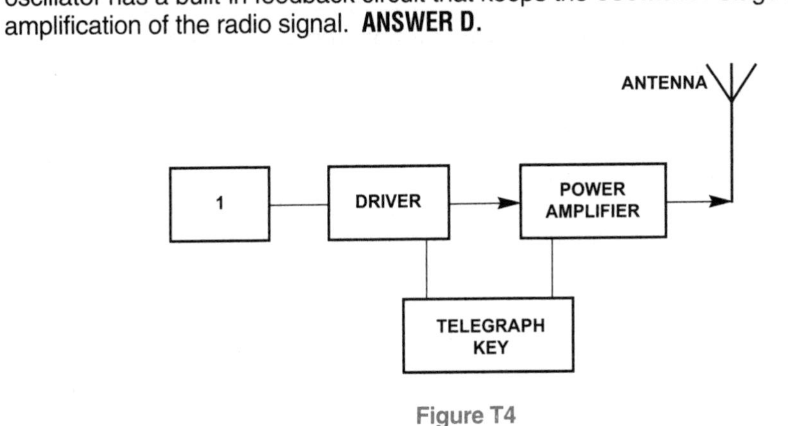

Figure T4

T4B10 Which of the following is an appropriate receive filter to select in order to minimize noise and interference for CW reception?

A. 500 Hz.
B. 1000 Hz.
C. 2400 Hz.
D. 5000 Hz.

For Morse code (CW), we restrict the bandwidth down to a skinny 400 Hz. Anything wider would bring in noise. Again, when you switch your worldwide radio to CW, the narrow ***500 Hz bandwidth filter*** automatically clicks in. **ANSWER A.**

T7A02 What type of receiver is shown in Figure T6?

A. Direct conversion.
B. Super-regenerative.
C. Single-conversion superheterodyne.
D. Dual-conversion superheterodyne.

This is a ***single-conversion superheterodyne*** receiver. You know it's single-conversion because it has only one intermediate frequency amplifier. **ANSWER C.**

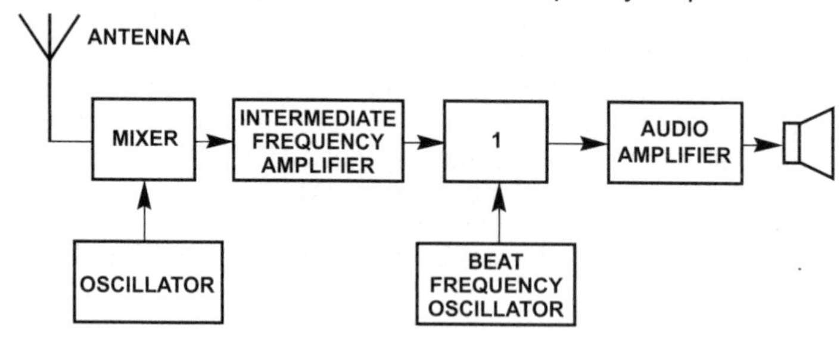

Figure T6

T7A01 What type of circuit does Figure T6 represent if block 1 is a product detector?

A. A simple phase modulation receiver.
B. A simple FM receiver.
C. A simple CW and SSB receiver.
D. A double-conversion multiplier.

A ***product detector*** is necessary in a simple Morse code ***(CW)*** and single-sideband ***(SSB)*** receiver. **ANSWER C.**

T8A01 Which of the following is a form of amplitude modulation?

A. Spread-spectrum.
B. Packet radio.
C. Single sideband.
D. Phase shift keying.

A form of ***amplitude modulation*** is called ***single sideband (SSB)***. With a single-sideband signal, the amplitude of the radio wave rises with the spoken word, up to a limit of around 2.8 kHz. Some hams employ double sideband, so the radio wave both rises and dips occupying almost twice as much bandwidth as single sideband. Single sideband is the type of modulation we use for weak-signal VHF and UHF work on Technician class frequencies. **ANSWER C.**

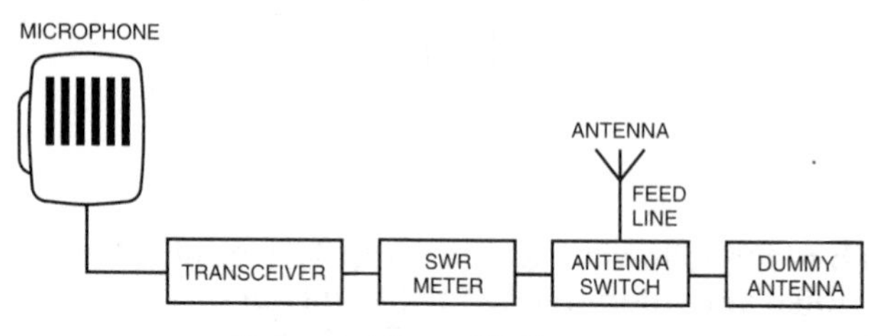

Voice or Phone Station

T8A08 What is the approximate bandwidth of a single sideband voice signal?

A. 1 kHz.
B. 3 kHz.
C. 6 kHz.
D. 15 kHz.

The properly-adjusted ***single sideband*** transmitter on VHF and UHF frequencies for Technician Class operators occupies approximately ***3 kHz of bandwidth***, depending on your voice characteristics. We employ upper sideband for VHF and UHF weak-signal work. Be sure to listen to my audio CD included with this book for the sounds of SSB. **ANSWER B.**

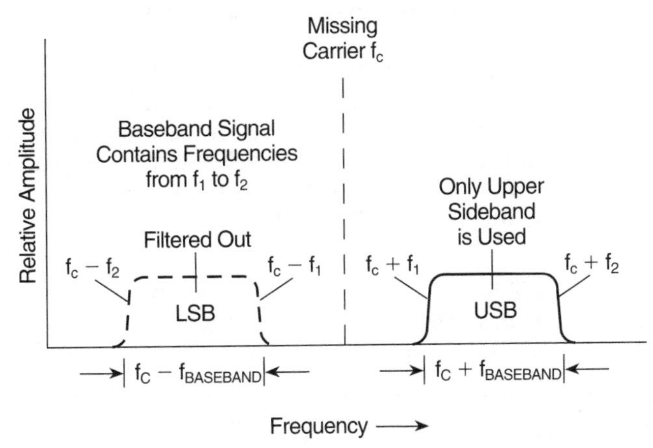

SSB signals are Amplitude Modulated (AM) with the carrier and one sideband suppressed.

T7A08 Which of the following circuits combines a speech signal and an RF carrier?

A. Beat frequency oscillator.
B. Discriminator.
C. Modulator.
D. Noise blanker.

Inside that brand new dual-band handheld is a circuit called the ***modulator***. It ***converts your spoken word*** from the microphone ***into a speech signal*** that is combined with the RF carrier. The modulator is always on the transmit side of your equipment, tied in to the microphone. **ANSWER C.**

T7A03 What is the function of a mixer in a superheterodyne receiver?

A. To reject signals outside of the desired passband.
B. To combine signals from several stations together.
C. To shift the incoming signal to an intermediate frequency.
D. To connect the receiver with an auxiliary device, such as a TNC.

The function of a ***mixer*** in a radio receiver is to ***shift the frequency*** of the receive signal so it can be processed by the ***intermediate frequency*** (IF) stages. **ANSWER C.**

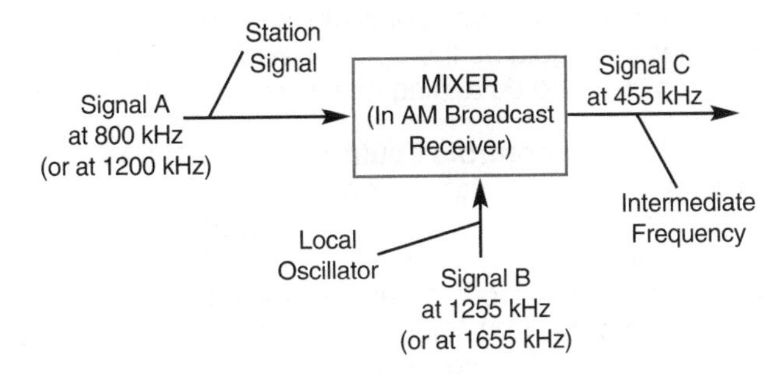

Block Diagram of an AM Broadcast Receiver Mixer
Source: *Basic Communications Electronics*,
© 1999 Master Publishing, Inc., Niles, IL

T4B08 What is the advantage of having multiple receive bandwidth choices on a multimode transceiver?

A. Permits monitoring several modes at once.
B. Permits noise or interference reduction by selecting a bandwidth matching the mode.
C. Increases the number of frequencies that can be stored in memory.
D. Increases the amount of offset between receive and transmit frequencies.

On 6 meters and 10 meters SSB and CW, the worldwide radio offers ***bandwidth choices***. When you change modes, your radio will automatically tighten up or expand the internal bandwidth filter selection. This allows you to ***minimize noise*** and maximize long range reception. **ANSWER B.**

HAM HINT: *Your first radio should be a two-band 2 meter/70 cm FM handheld. The handheld keeps you in touch, locally. You also get some out-of-state and out-of-country DX, too, when local repeaters tie into the internet using voice over Internet protocol (VOIP).*

But, if you're really serious about exploring SSB satellite calls or moon bounce, meteor scatter, aurora, and 6 meter and 10 meter voice skip, plus all the fun of CW on high frequency Technician sub-bands, then consider a worldwide, high-frequency transceiver that includes 6 meters, 2 meters and 70 cm.

Not many HF transceivers give you these higher bands – sometimes they may give you 6 meters, but nothing higher. Each manufacturer has one or two specialty HF transceivers that give you the VHF and UHF bands, as well! They run about $900 to $1,400, but what you get is a worldwide DX transceiver, plus multimode capability on 6 meters, 2 meters, and 70 cm too! Best of all, everything is built into one nice neat, compact unit – nothing else to add other than your higher-band VHF and UHF antenna systems.

These are 12 volt radios that can be operated mobile as well as from a power supply at the house. This is what I run in our communications van, and it's a great way to have a single radio that will serve you well all the way up to your Extra Class license! Oh yeah, they give you 2 meter and 70 cm FM, too!

T4B09 Which of the following is an appropriate receive filter to select in order to minimize noise and interference for SSB reception?

A. 500 Hz.
B. 1000 Hz.
C. 2400 Hz.
D. 5000 Hz.

For ***single sideband***, the normal voice bandwidth is about 2500 Hz, so the ***2400 Hz*** bandwidth selection would be the appropriate filter. This usually happens automatically, so you don't need to do a thing to enjoy weak signal work! **ANSWER C.**

T4B06 Which of the following controls could be used if the voice pitch of a single-sideband signal seems too high or low?

A. The AGC or limiter.
B. The bandwidth selection.
C. The tone squelch.
D. The receiver RIT or clarifier.

If you plan to do some weak signal work, great! We need more weak signal operators using CW (Morse code) and single sideband. It will take a little practice to tune in voice properly on single sideband. You should set your ***receiver RIT or clarifier***, to the neutral position, and then adjust slightly the big tuning knob ***for proper SSB voice reception***. You can then make fine adjustments with the receiver RIT or clarifier. **ANSWER D.**

T4B07 What does the term "RIT" mean?

A. Receiver Input Tone.
B. Receiver Incremental Tuning.
C. Rectifier Inverter Test.
D. Remote Input Transmitter.

The "RIT" control is only found on high-end, weak-signal, mobile and base station equipment capable of SSB. The ***Receiver Incremental Tuning control*** allows you to tune the receiver slightly up and down about a kilohertz without changing your transmit frequency, which is useful for working satellite Doppler shift. For non-satellite work, leave the RIT control disabled so you are always transmitting and receiving on your same frequency. **ANSWER B.**

Receiver section in a communications transceiver

T7A12 Where is an RF preamplifier installed?

A. Between the antenna and receiver.
B. At the output of the transmitter's power amplifier.
C. Between a transmitter and antenna tuner.
D. At the receiver's audio output.

Don't go out and buy one. Your radio equipment already has an excellent built-in preamplifier. Only if you purchased a very old radio, with noticeably weak reception, would you consider an external preamplifier. The ***pre-amp goes in between the antenna input and the receiver***. Be careful how you wire it in – if you simply put it on the outside of the equipment, the first time you transmit the pre-amp will be history! It either needs external switching, or must be placed inside the equipment, between the antenna line and the receiver section. **ANSWER A.**

T7B02 What is meant by fundamental overload in reference to a receiver?

A. Too much voltage from the power supply.
B. Too much current from the power supply.
C. Interference caused by very strong signals.
D. Interference caused by turning the volume up too high.

When someone nearby is pumping out a really strong signal, and you're on your handheld trying to contact that distant repeater, your ***radio will become overloaded with interference caused by that very strong signal***. **ANSWER C.**

T7A13 Which term describes the ability of a receiver to discriminate between multiple signals?

A. Tuning rate.
B. Sensitivity.
C. Selectivity.
D. Noise floor.

The capability of any radio equipment to hear specifically one frequency, and ***discriminate*** against ***signals*** on either side of that frequency, is called selectivity. On VHF and UHF equipment, the selectivity is usually pre-set, and most equipment is plenty selective. On the worldwide radio gear, you can sometimes add more filters for increased ***selectivity***. While additional filters may cut down on the fidelity of the received signal, you can "tighten up" on the receiver response. **ANSWER C.**

T2B05 What determines the amount of deviation of an FM signal?

A. Both the frequency and amplitude of the modulating signal.
B. The frequency of the modulating signal.
C. The amplitude of the modulating signal.
D. The relative phase of the modulating signal and the carrier.

How wide will your FM signal be on the air? That's determined by the amplitude (height) of your modulating signal – your voice. The amplitude of the modulating signal is how strongly you are speaking into your transceiver, which will determine the amount of ***carrier deviation***. Sounds complicated, but after a while you'll discover that you need to "close talk" your radio microphone or dual-band handheld in order to get the proper amount of ***amplitude*** for your modulating signal. **ANSWER C.**

T2B06 What happens when the deviation of an FM transmitter is increased?

A. Its signal occupies more bandwidth.
B. Its output power increases.
C. Its output power and bandwidth increases.
D. Asymmetric modulation occurs.

You're going to hear about areas of the US where all repeaters have been "narrow banded," allowing a modulation deviation of no greater than plus or minus 2.5 kHz. If you run regular deviation, plus or minus 5 kHz, this will ***occupy more bandwidth***, and that local narrow band repeater will not properly pass your transmitted signal, unless you speak very softly! **ANSWER A.**

T7A04 What circuit is pictured in Figure T7, if block 1 is a frequency discriminator?

A. A double-conversion receiver.
B. A regenerative receiver.
C. A superheterodyne receiver.
D. An FM receiver.

Did you spot the words "***frequency discriminator***" in this question? This means we are looking at a ***frequency modulation (FM) receiver***. **ANSWER D.**

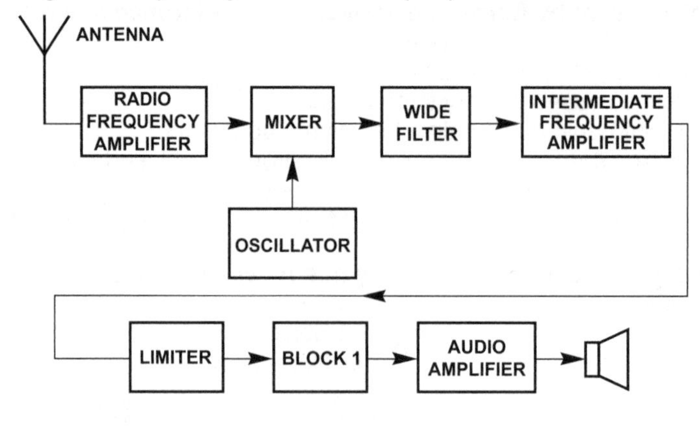

Figure T7

T7A11 Which of the following circuits demodulates FM signals?

A. Limiter.
B. Discriminator.
C. Product detector.
D. Phase inverter.

When we receive signals, it takes place in your handheld's discriminator section. FM signals are ***demodulated*** from the incoming carrier, and ultimately passed on to the audio stage that feeds you handheld speaker. ***Discriminator*** – FM receiver component. **ANSWER B.**

T8A10 What is the typical bandwidth of analog fast-scan TV transmissions on the 70 cm band?

A. More than 10 MHz.
B. About 6 MHz.
C. About 3 MHz.
D. About 1 MHz.

A fascinating type of ham radio operation is called fast-scan television, abbreviated "ATV." Hams can send fast-scan television pictures direct or through special fast-scan television repeaters over hundreds of miles! Years ago, the Pasadena Rose Parade was covered with fast-scan TV pictures, keeping officials up-to-date on float safety. Fast-scan television is very wide – ***about 6 MHz*** – and can be found on 70-centimeter frequencies and higher where equipment cost is modest, especially if you already own a video camera. **ANSWER B.**

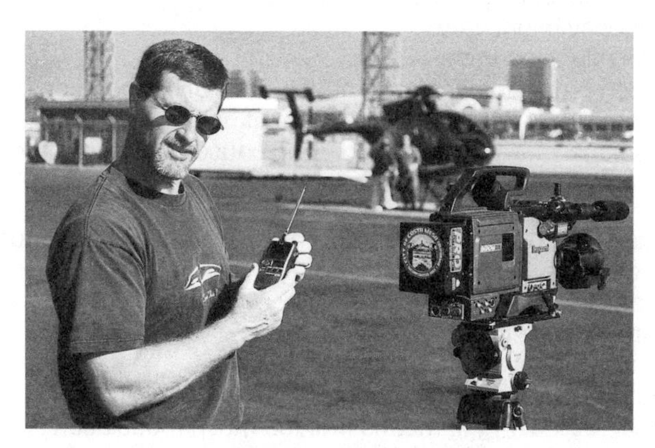

Amateur TV signals can be received on a variety of equipment – even a small hand-held monitor.

Run Some Interference Protection

T7B10 What might be the problem if you receive a report that your audio signal through the repeater is distorted or unintelligible?

A. Your transmitter may be slightly off frequency.
B. Your batteries may be running low.
C. You could be in a bad location.
D. All of these choices are correct.

This particular question about a poor signal report from your radio through a repeater has a correct answer showing ***all of the conditions*** that might lead to this bad reception. Likely, a bad location in one of those "not spots" could lead to a distorted or weak signal, so try moving to a slightly different location where the signal strength of the repeater shows greater on your handheld LCD graph, and that should probably be a better spot to transmit through the repeater. **ANSWER D.**

T4B01 What may happen if a transmitter is operated with the microphone gain set too high?

A. The output power might be too high.
B. The output signal might become distorted.
C. The frequency might vary.
D. The SWR might increase.

Some base station radios for weak signal VHF and UHF operation have a microphone gain control. Set the control at about half scale. Turning the control wide open will set the ***mic gain too high*** and cause the transmitted ***signal*** to become ***distorted***. **ANSWER B.**

T7B01 What can you do if you are told your FM handheld or mobile transceiver is over deviating?

A. Talk louder into the microphone.
B. Let the transceiver cool off.
C. Change to a higher power level.
D. Talk farther away from the microphone.

If your set is ***over-deviating***, it means that too much modulation is driving your signal beyond its normal bandwidth. If you ***talk farther away from the microphone***, you will minimize or even eliminate the over-deviation. **ANSWER D.**

T2B07 What should you do if you receive a report that your station's transmissions are causing splatter or interference on nearby frequencies?

A. Increase transmit power.
B. Change mode of transmission.
C. Report the interference to the equipment manufacturer.
D. Check your transmitter for off-frequency operation or spurious emissions.

Everything is fine with your new dual-band handheld until that fateful day you drove off with it sitting on the roof of your car. Luckily, your better half scraped it up from the driveway, and when you returned home you were happy to see that it still turns on and appears to transmit. However, every time you transmit, other hams now say your scratched-up handheld is sounding ***off frequency***, and it is sending out ***spurious emissions***, which are multiple signals on other nearby frequencies. More than likely, the drop to the concrete driveway fractured internal coil "slugs" inside the handheld, throwing off its precise tuning. You will likely need to send the handheld to a service center, explaining that it did a swan dive from your roof to the driveway, and this will give them a heads up to look for the tiny coil element slugs that need to be replaced. **ANSWER D.**

T4B05 Which of the following would reduce ignition interference to a receiver?

A. Change frequency slightly.
B. Decrease the squelch setting.
C. Turn on the noise blanker.
D. Use the RIT control.

On that multi-mode, weak-signal radio when you receive a weak single-sideband signal, turn on the ***noise blanker to minimize ignition noise***. There is no noise blanker function on a common FM handheld or mobile radio, just on your bigger high-frequency, multi-mode transceiver. **ANSWER C.**

T7B09 What could be happening if another operator reports a variable high-pitched whine on the audio from your mobile transmitter?

A. Your microphone is picking up noise from an open window.
B. You have the volume on your receiver set too high.
C. You need to adjust your squelch control.
D. Noise on the vehicle's electrical system is being transmitted along with your speech audio.

This is a common problem with a handheld plugged into the 12 volt accessory socket in your automobile. If your automobile has a hefty alternator charging system without filters on the leads, the ***AC from the alternator*** will put a high pitch whine on your transmit and receive signal. If you hear a whine that changes as you accelerate and then comes back down in pitch as you let off the gas, plan to buy some DC noise filters and put them on both the electrical system as well as on your handheld 12-volt DC input cord. Most car adaptor plugs for handhelds have built-in noise chokes to minimize this common problem. **ANSWER D.**

T4A10 What is the source of a high-pitched whine that varies with engine speed in a mobile transceiver's receive audio?

A. The ignition system.
B. The alternator.
C. The electric fuel pump.
D. Anti-lock braking system controllers.

You have that brand new mobile radio installed in your brand new car (you rewarded yourself for passing the exam), but each time you goose the accelerator you hear a high-pitched whine coming out of your mobile radio speaker! This is most likely ***alternator whine***, and can be minimized with DC filter chokes and alternator filters. **ANSWER B.**

T4A09 Which would you use to reduce RF current flowing on the shield of an audio cable?

A. Band-pass filter.
B. Low-pass filter.
C. Preamplifier.
D. Ferrite choke.

If you plan to go digital with your new Technician Class privileges, load up on a handful of ferrite chokes. These are clamshell iron devices that simply snap on over the wiring coming from your handheld or mobile radio to your computer. The ***ferrite choke will minimize RF*** (radio frequency) currents flowing ***on the shield*** of an audio cable. You want everything on the inside, not the outside! **ANSWER D.**

T4A05 What type of filter should be connected to a TV receiver as the first step in trying to prevent RF overload from a nearby 2 meter transmitter?

A. Low-pass filter.
B. High-pass filter.
C. Band-pass filter.
D. Band-reject filter.

Now that over-the-air television broadcasts have gone from analog to all digital, ham RF overload to a TV receiver is a rare occurrence. Most internal digital converters – plus almost all digital external converters – now receive local television signals on UHF frequencies, well above the 2 meter band. However, if your 2-meter base station ever should interfere with over-the-air TV reception, a ***band reject filter*** could help solve the problem. But I wouldn't worry about this just yet, because you are still studying for the exam, and most of you will start out with a dual-band handheld that will likely never interfere with TV digital reception. **ANSWER D.**

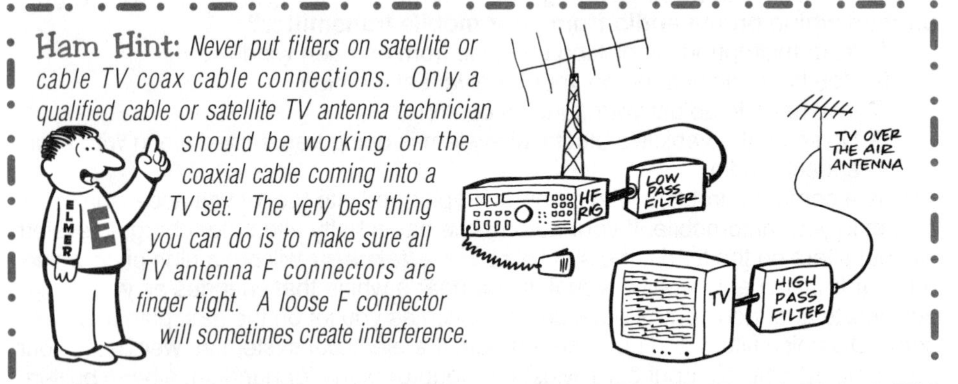

Ham Hint: *Never put filters on satellite or cable TV coax cable connections. Only a qualified cable or satellite TV antenna technician should be working on the coaxial cable coming into a TV set. The very best thing you can do is to make sure all TV antenna F connectors are finger tight. A loose F connector will sometimes create interference.*

T4A04 Where must a filter be installed to reduce harmonic emissions?

A. Between the transmitter and the antenna.
B. Between the receiver and the transmitter.
C. At the station power supply.
D. At the microphone.

Now that television signals are digital, and the majority are now transmitted over-the-air on UHF frequencies, ham radio television interference (TVI) is a rarity. Your worldwide radio already has a low pass filter installed to remove harmonics. On the VHF and UHF ham bands, no filter should ever be added to your antenna system. But to answer this question correctly, if you decided to add a ***low pass filter*** on the antenna connection of your worldwide radio, it goes between ***the transmitter and the antenna***. **ANSWER A.**

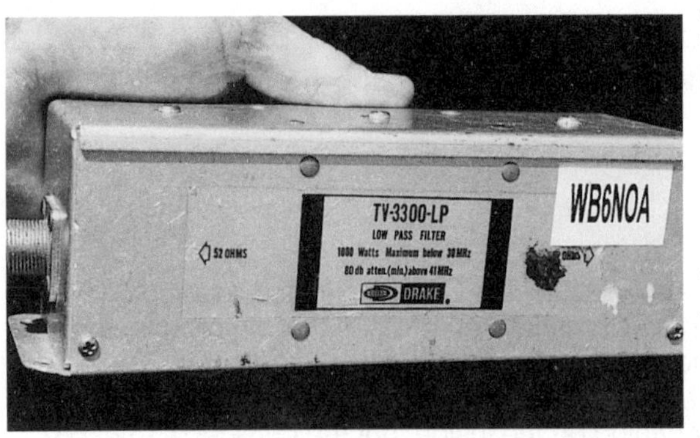

There are low-pass filters like this one, band-pass filters, and high-pass filters that can be used to solve interference problems.

T7B03 Which of the following may be a cause of radio frequency interference?

A. Fundamental overload.
B. Harmonics.
C. Spurious emissions.
D. All of these choices are correct.

You are outside sitting in your deck chair listening to your local repeater when, all of a sudden, your handheld picks up nothing but hash. Your buddy across the street (also a ham) is transmitting with a mobile radio and amplifier that he bought off a swap net that was advertised as "may need some tweaking." That nearby transmitter is probably ***overloading your receiver with harmonics and spurious emissions, and all of this leads to radio frequency interference***. Don't buy questionable used gear – Marconi himself probably could not fix it.
ANSWER D.

T7B11 What is a symptom of RF feedback in a transmitter or transceiver?

A. Excessive SWR at the antenna connection.
B. The transmitter will not stay on the desired frequency.
C. Reports of garbled, distorted, or unintelligible transmissions.
D. Frequent blowing of power supply fuses.

You are on 6 meters for the first time, and signal strength tests with hams a few miles away indicate you have a strong signal, but your modulation ***(your voice) is slightly garbled***. This is most likely caused by RF feedback between your antenna system and the microphone. You have the antenna temporarily set on a tripod, just a few feet away from your operating station. Get some longer coax; move the antenna at least 15 feet away from your operating station and likely your friends will tell you the distortion has cleared up. No more ***RF feedback***!
ANSWER C.

T7B06 What should you do first if someone tells you that your station's transmissions are interfering with their radio or TV reception?

A. Make sure that your station is functioning properly and that it does not cause interference to your own television.
B. Immediately turn off your transmitter and contact the nearest FCC office for assistance.
C. Tell them that your license gives you the right to transmit and nothing can be done to reduce the interference.
D. Continue operating normally because your equipment cannot possibly cause any interference.

It's unlikely that VHF and UHF signals will cause television interference to your neighbors on an outside antenna, or on the dish or on cable. However, if you go on the worldwide bands using your Morse code privileges, be aware that these frequencies could sneak into a television set and cause problems. First, double check that your equipment is well grounded and operating properly, and ***double check that your TV is working okay*** when you are transmitting over the air. If it is, chances are your neighbors may have some loose connections on their TV receivers, and this should be cured as a step to reducing the interference from your station when transmitting on high frequency. **ANSWER A.**

T7B04 What is the most likely cause of interference to a non-cordless telephone from a nearby transmitter?

A. Harmonics from the transmitter.
B. The telephone is inadvertently acting as a radio receiver.
C. Poor station grounding.
D. Improper transmitter adjustment.

Inexpensive corded telephones will act like a tiny ***radio receiver*** to your transmitter's signals if your handheld is only a foot away. Get some separation by moving away from the telephone and the interference should go away. **ANSWER B.**

T7B05 What is a logical first step when attempting to cure a radio frequency interference problem in a nearby telephone?

A. Install a low-pass filter at the transmitter.
B. Install a high-pass filter at the transmitter.
C. Install an RF filter at the telephone.
D. Improve station grounding.

You can purchase snap-on ferrite chokes that act as an ***RF filter at the telephone*** equipment. Snap these filters over the telephone power cord, curly cord, and on the actual incoming telephone line cord. The more snap-on filters you add, the less likely your little handheld or base station will interfere. **ANSWER C.**

T7B07 Which of the following may be useful in correcting a radio frequency interference problem?

A. Snap-on ferrite chokes.
B. Low-pass and high-pass filters.
C. Band-reject and band-pass filters.
D. All of these choices are correct.

The best way to correct radio frequency interference from your ham radio to your mother-in-law's HI-FI and ancient black-and-white analog TV would be to use snap-on ferrite chokes, and rely on your own equipment's band pass filters to keep your transmit signal clean. Hopefully, that old TV set has its own high pass filter built in. For the worldwide ham bands, newer ham rigs have low pass filters built in and, in rare cases, you could actually build a "shorted stub" band reject filter for that old TV set. For the exam, ***all of these choices are correct***, but in the real world of ham radio, the very best way of eliminating your lovely voice coming over you mom-in-law's hair dryer and toaster would be snap-on ferrite chokes, especially useful on computer leads, telephone modems, hi-fi connections, and on incoming telephone lines and the telephone handset. A good ham has a 5 pound bag of ferrite chokes ready for any interference problem! But for the exam, go with all of the answers! **ANSWER D.**

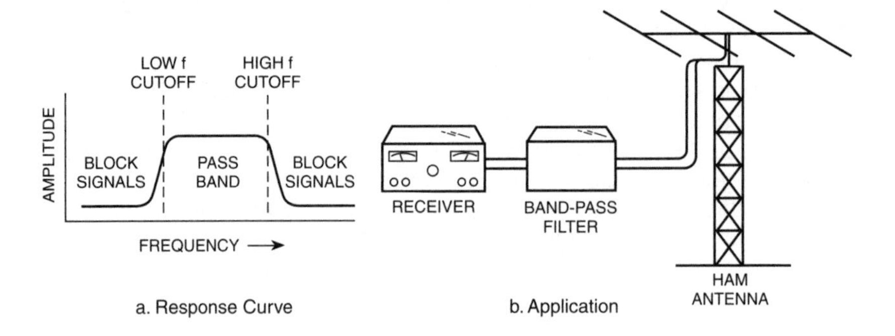

Band-Pass Filter

T7B08 What should you do if a "Part 15" device in your neighbor's home is causing harmful interference to your amateur station?

A. Work with your neighbor to identify the offending device.
B. Politely inform your neighbor about the rules that require him to stop using the device if it causes interference.
C. Check your station and make sure it meets the standards of good amateur practice.
D. All of these choices are correct.

A wireless ANYTHING is likely rated a Part 15 device, and this means the potential of ham radio interference. My neighbor's wireless weather station sometimes clobbers my weak signal work on 432 MHz. And when I transmit moonbounce, another neighbor reports his wireless doorbell goes bing-bong. As more wireless Part 15 devices switch from analog to digital, our ham transmitters probably won't disturb them anymore. However, it's good to take stock of what you may have around the house that is a Part 15 device, and if you have a pesky weird sound on a certain 2 meter frequency, see if that particular device is causing it. ***All of these*** Part 15 devices going digital is good for minimizing ham radio interference! **ANSWER D.**

A simple snap-on choke filter like this one can help resolve harmful interference problems on Part 15 devices.

Ham Hint: *A Part 15 device is any flea-powered transmitter found all over your neighborhood.*

Wireless routers	*Wireless weather stations*
Wireless cordless phones	*Wireless temperature swimming pool probe*
Wireless headphones	*Wireless printers and FAX machines*

Electrons- Go With The Flow!

T5A05 What is the electrical term for the electromotive force (EMF) that causes electron flow?

A. Voltage.
B. Ampere-hours.
C. Capacitance.
D. Inductance.

Think of voltage as water pressure in your kitchen plumbing! Open the valve (turn on the switch) and ***current*** begins to ***flow***. The pressure is ***voltage***, and the trickle of water could be similar to current. **ANSWER A.**

T5A11 What is the basic unit of electromotive force?

A. The volt.
B. The watt.
C. The ampere.
D. The ohm.

Another name for voltage is electromotive force – "may the force be with you" – voltage. ***The volt***. **ANSWER A.**

T7D01 Which instrument would you use to measure electric potential or electromotive force?

A. An ammeter.
B. A voltmeter.
C. A wavemeter.
D. An ohmmeter.

Another name for electromotive force is voltage. We ***measure voltage*** with a ***voltmeter***. **ANSWER B.**

T7D02 What is the correct way to connect a voltmeter to a circuit?

A. In series with the circuit.
B. In parallel with the circuit.
C. In quadrature with the circuit.
D. In phase with the circuit.

We ***test for voltage*** by hooking our meter across the voltage source without undoing any wires – ***a parallel connection***. Checking the voltage when operating equipment is called "checking the voltage source under load." **ANSWER B.**

T6A10 What is the nominal voltage of a fully charged nickel-cadmium cell?

A. 1.0 volts.
B. 1.2 volts.
C. 1.5 volts.
D. 2.2 volts.

A brand new AA alkaline battery rests at 1.5 volts. But in this question, they ask about the rechargeable nickel-cadmium battery that usually rests at about 1.2 volts. Your handheld radio won't know the difference! Just remember, ***1.2 volts for a fully charged nickel-cadmium battery***. **ANSWER B.**

Ni-Cad rechargeable 1.25 volt batteries in a marine hand held.

T6A11 Which battery type is not rechargeable?

A. Nickel-cadmium.
B. Carbon-zinc.
C. Lead-acid.
D. Lithium-ion.

Luckily you won't find any handheld shipped with ***carbon-zinc, non-rechargeable batteries***. Alkaline and carbon-zinc batteries only work once and should be properly disposed of when depleted. **ANSWER B.**

T5A06 How much voltage does a mobile transceiver usually require?

A. About 12 volts.
B. About 30 volts.
C. About 120 volts.
D. About 240 volts.

Most mobile transceivers run off ***12 volts*** DC. If you are purchasing a handheld, be sure to buy the inexpensive 12 volt DC adaptor. This plugs into the accessory socket in your vehicle and will nicely power a handheld. Do NOT use the accessory socket to power anything more than just you little handheld! OK, you can power your cell phone, too. **ANSWER A.**

T4A11 Where should a mobile transceiver's power negative connection be made?

A. At the battery or engine block ground strap.
B. At the antenna mount.
C. To any metal part of the vehicle.
D. Through the transceiver's mounting bracket.

Ham radio power leads need to be connected directly at the battery source. This means positive to the battery positive terminal and negative to the ***battery negative terminal or*** the nearby ***engine block ground strap***. This will minimize alternator whines and whistles! **ANSWER A.**

T5A03 What is the name for the flow of electrons in an electric circuit?

A. Voltage.
B. Resistance.
C. Capacitance.
D. Current.

Think of the flow of electrons as the flow of water in a stream. If you get out there in midstream, you will feel the ***current***. **ANSWER D.**

T7D04 Which instrument is used to measure electric current?

A. An ohmmeter.
B. A wavemeter.
C. A voltmeter.
D. An ammeter.

Current is measured in amperes, the unit of current. We use an ***ammeter*** to measure electrical current. **ANSWER D.**

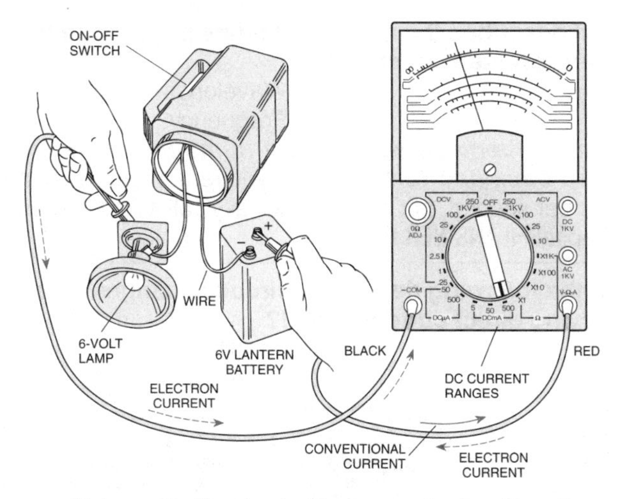

Using a Multimeter to Measure a Series Circuit

Source: *Basic Electronics* © 1994, 2000, Master Publishing, Inc., Niles, Illinois

T7D03 How is an ammeter usually connected to a circuit?

A. In series with the circuit.
B. In parallel with the circuit.
C. In quadrature with the circuit.
D. In phase with the circuit.

An ammeter measures current. To measure current, turn off the power, disconnect one lead of the load from its source voltage (for example, at the fuse holder) and insert a ***ammeter in series*** with that lead. If you are measuring DC current, you will need to connect the meter with the correct polarity, so the meter reads up scale when power is turned on. **ANSWER A.**

T5A01 Electrical current is measured in which of the following units?

A. Volts.
B. Watts.
C. Ohms.
D. Amperes.

The flow of electrons in a conductor is called current. Current is measured in ***amperes***. Amperes is often referred to as "amps." **ANSWER D.**

T5A07 Which of the following is a good electrical conductor?

A. Glass.
B. Wood.
C. Copper.
D. Rubber.

Most wire is ***copper***, and this is a good conductor. Some relays use gold- or silver-plated contacts, and these are great conductors! You can use aluminum foil as a ground plane; it also is a good conductor. Glass, wood and rubber are insulators! **ANSWER C.**

T5A09 What is the name for a current that reverses direction on a regular basis?

A. Alternating current.
B. Direct current.
C. Circular current.
D. Vertical current.

Have you ever been bitten by a hot power cord? Think back to that shocking time, and recall your accidental zap feeling much like a buzz. It is exactly that – ***alternating current*** reversing directions 60 times a second. So beside the shock, that tingling buzz is the feeling of alternating current. **ANSWER A.**

T3B02 What term describes the number of times per second that an alternating current reverses direction?

A. Pulse rate.
B. Speed.
C. Wavelength.
D. Frequency.

When we measure ***frequencies***, we count the ***number of times per second that current flows back and forth***. And do you remember the name of the word frequency? Cycles per second, but now we officially call it hertz, abbreviated Hz. Don't forget the capital H! **ANSWER D.**

T6D01 Which of the following devices or circuits changes an alternating current into a varying direct current signal?

A. Transformer.
B. Rectifier.
C. Amplifier.
D. Reflector.

Most ham equipment works off of DC voltage. The way we transform household AC power over to DC is with a rectifier circuit. The rectifier uses

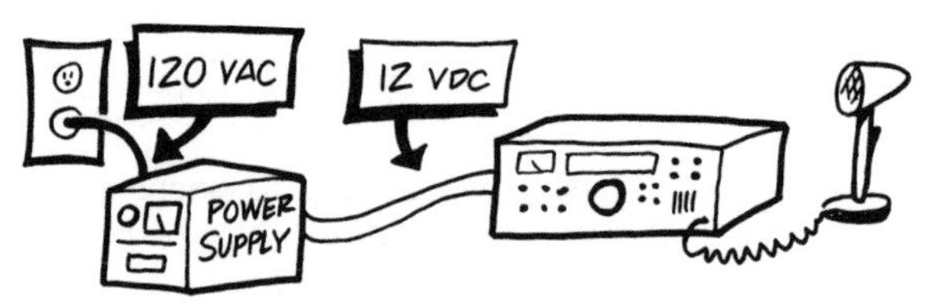

diodes to block the full house power AC cycle, leading to a varying direct current signal. Capacitors are used to filter and smooth out this fluctuating DC current. Thanks to the *rectifier* circuit, we can *change AC into varying DC*. **ANSWER B.**

T5A04 What is the name for a current that flows only in one direction?

A. Alternating current.
B. Direct current.
C. Normal current.
D. Smooth current.

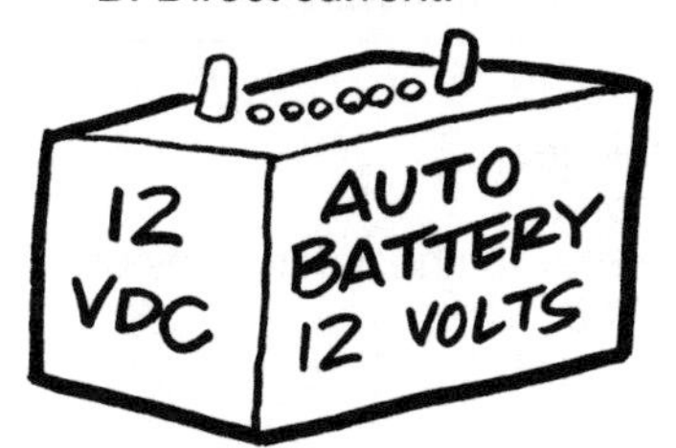

Batteries generate *direct current*. Even though a current may vary in value, if it always flows in the same direction, it is a direct current (DC). **ANSWER B.**

T6B02 What electronic component allows current to flow in only one direction?

A. Resistor.
B. Fuse.
C. Diode.
D. Driven Element.

The process of changing AC to pulsating DC is called rectification. Big word, huh? We use a diode in a circuit to allow current to flow in only one direction. *The diode* – much like a check valve – *stops current flow when it tries to go in the reverse direction*. **ANSWER C.**

T6B09 What are the names of the two electrodes of a diode?

A. Plus and minus.
B. Source and drain.
C. Anode and cathode.
D. Gate and base.

First take a look at the symbol for a diode. The cathode is represented by the short straight line at the tip of the arrow and the anode is represented by the arrow itself. Now think backward – current flows in the opposite direction of the arrow from cathode to anode, forward bias. There is almost no current flow in the direction of the arrow from *anode* to *cathode*. In a silicon diode, it takes about a half volt for conduction. **ANSWER C.**

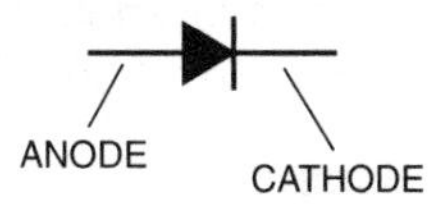

Here is the schematic symbol of a diode. Current will only flow ONE WAY in a diode. You can remember this diode diagram as a one-way arrow (key words).

Semiconductor Diode

ANODE CATHODE

Here is the schematic symbol of a Zener diode. Since a diode only passes energy in one direction, look for that one-way arrow, plus a "Z" indicating it is a Zener diode. Doesn't that vertical line look like a tiny "Z"?

Zener Diode

T6B06 How is a semiconductor cathode lead usually identified?

A. With the word "cathode."
B. With a stripe.
C. With the letter "C."
D. All of these choices are correct.

Not much room on the tiny diode to put a letter or a word to indicate the *cathode* end, so a *tiny stripe* does the trick nicely. **ANSWER B.**

T6A01 What electrical component is used to oppose the flow of current in a DC circuit?

A. Inductor.
B. Resistor.
C. Voltmeter.
D. Transformer.

The ***opposition to the flow of current*** in a direct current circuit is called resistance. Just like those beavers blocking up the current flow in a stream. The unit of resistance in a DC circuit is the Ohm. The component is called a ***resistor***. **ANSWER B.**

T7D05 What instrument is used to measure resistance?

A. An oscilloscope.
B. A spectrum analyzer.
C. A noise bridge.
D. An ohmmeter.

We use an ***ohmmeter*** to check for ohms of resistance. **ANSWER D.**

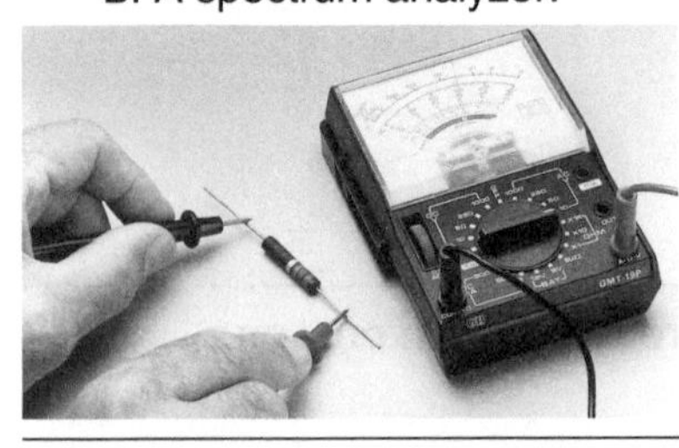

A D'Arsonval-type meter uses a mechanical needle to indicate the test result

T6A02 What type of component is often used as an adjustable volume control?

A. Fixed resistor.
B. Power resistor.
C. Potentiometer.
D. Transformer.

Ham operators are about the last to enjoy the potentiometer. This is like an old fashioned rheostat, where a wiper brush makes contact with wire windings, changing the variable resistance. This is found in the ***volume control*** of most handheld ham radios. That new plasma TV doesn't have a ***potentiometer***, rather volume is controlled by a digital push button circuit. In ham sets, we usually have the beloved potentiometer! **ANSWER C.**

T6A03 What electrical parameter is controlled by a potentiometer?

A. Inductance.
B. Resistance.
C. Capacitance.
D. Field strength.

The ***potentiometer varies the resistance*** in most ham radio volume control circuits. If your ham set does not have a round volume control knob, but only push buttons for volume control, you don't have a potentiometer any more. **ANSWER B.**

T5A08 Which of the following is a good electrical insulator?

A. Copper.
B. Glass.
C. Aluminum.
D. Mercury.

Old-time hams will lead you down to their basement to show off their collection of multi-color ***glass*** power pole insulators. Clean glass is a great insulator, and generally won't conduct electrons unless it gets covered with dirt and ocean salt air. Power pole maintenance workers will use a special non-conductive water jet to clean the high-voltage insulators. **ANSWER B.**

T6A06 What type of electrical component stores energy in a magnetic field?

A. Resistor.
B. Capacitor.
C. Inductor.
D. Diode.

It is the coil, called an ***inductor***, that ***stores energy in the magnetic field***. We can actually see the effects of an inductor with a magnetic compass. **ANSWER C.**

T6A07 What electrical component is usually composed of a coil of wire?

A. Switch.
B. Capacitor.
C. Diode.
D. Inductor.

The ***coil*** of wire is called an ***inductor***. **ANSWER D.**

T5C03 What is the ability to store energy in a magnetic field called?

A. Admittance.
B. Capacitance.
C. Resistance.
D. Inductance.

Remember that coils develop a magnetic field, which can be detected by holding a magnetic compass held near the energized coil. ***Energy stored in a magnetic field is called inductance***. **ANSWER D.**

T5C04 What is the basic unit of inductance?

A. The coulomb.
B. The farad.
C. The henry.
D. The ohm.

The basic unit of ***inductance*** is the ***henry***. Because the henry is a fairly large unit, we usually measure inductance in one thousandths of a henry (millihenry) and one millionths of a henry (microhenry). **ANSWER C.**

T5C01 What is the ability to store energy in an electric field called?

A. Inductance.
B. Resistance.
C. Tolerance.
D. Capacitance.

Capacitors store ***energy in an electric field***. This is called ***capacitance***. **ANSWER D.**

T5C02 What is the basic unit of capacitance?

A. The farad.
B. The ohm.
C. The volt.
D. The henry.

The basic unit of ***capacitance*** is the ***farad***. Because the farad is a fairly large unit, we usually measure capacitance in one millionths of a farad (microfarad) or one million millionths of a farad (picofarad). **ANSWER A.**

T6A04 What electrical component stores energy in an electric field?

A. Resistor.
B. Capacitor.
C. Inductor.
D. Diode.

It is the capacitor that stores energy in an electric field. Got it? ***Electric field is the capacitor***. **ANSWER B.**

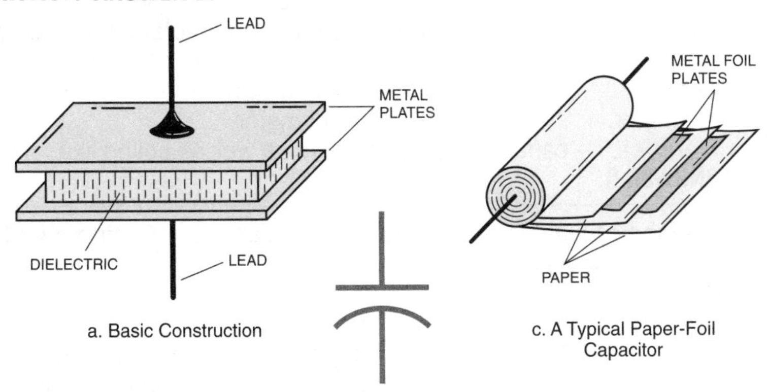

Typical construction and schematic symbol for capacitors.

T6A05 What type of electrical component consists of two or more conductive surfaces separated by an insulator?

A. Resistor.
B. Potentiometer.
C. Oscillator.
D. Capacitor.

The ***capacitor has 2 or more conductive surfaces, separated by insulation***. When a capacitor shorts out or breaks down, it's usually an old one, where the insulation has finally dried up, giving you all sorts of noise as a leaky older capacitor. **ANSWER D.**

T6A08 What electrical component is used to connect or disconnect electrical circuits?

A. Zener diode.
B. Switch.
C. Inductor.
D. Variable resistor.

The common ***switch*** is that component which connects or interrupts an electrical circuit's continuity. If the switch is open, the circuit usually will not function. Close the switch and, presto, your radio comes to life. Your new dual-band handheld turns on with either a volume control "click" or a keypad pushbutton. If any handheld turns on and off with a volume control "click," this switch totally turns off any battery drain to the inside electronics when you turn the radio off. However, many handhelds have a "soft" pushbutton turn on, and this type of transistorized electronic switch always continues to draw a fraction of a milliamp when turned off! A "soft" turn-on switch may ultimately drain your radio's battery over a month's time, even when the equipment is off. Simply remove the battery if you plan not to use your gear for several months. **ANSWER B.**

T6A09 What electrical component is used to protect other circuit components from current overloads?

A. Fuse.
B. Capacitor.
C. Shield.
D. Inductor.

We use a ***fuse to protect components*** from excessive amounts of current flow. Even handheld radio batteries have fuses, in case you take that freshly charged battery and stick it in your pocket with a lot of spare change. The battery shorts out, you get a burning sensation in your thigh, but luckily the fuse opens, and all you get is a dead battery and a lot of hot quarters. Some batteries use thermal fuses, which automatically reset once they cool down. **ANSWER A.**

T6B03 Which of these components can be used as an electronic switch or amplifier?

A. Oscillator.
B. Potentiometer.
C. Transistor.
D. Voltmeter.

Here is the ***transistor*** – capable of amplification, as well as acting like an on and off switch. **ANSWER C.**

T6B01 What class of electronic components is capable of using a voltage or current signal to control current flow?

A. Capacitors.
B. Inductors.
C. Resistors.
D. Transistors.

Most ham sets don't use tubes anymore. No tubes in that little dual band handheld. All transistors! ***Transistors*** are tiny electrical components capable of using a tiny voltage or current signal to control larger amounts of current flow to other circuits. **ANSWER D.**

TRANSISTOR BASICS

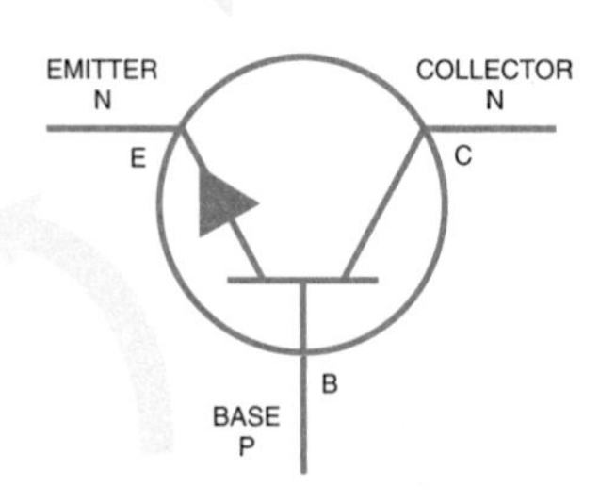

a. Schematic Symbol of NPN Transistor

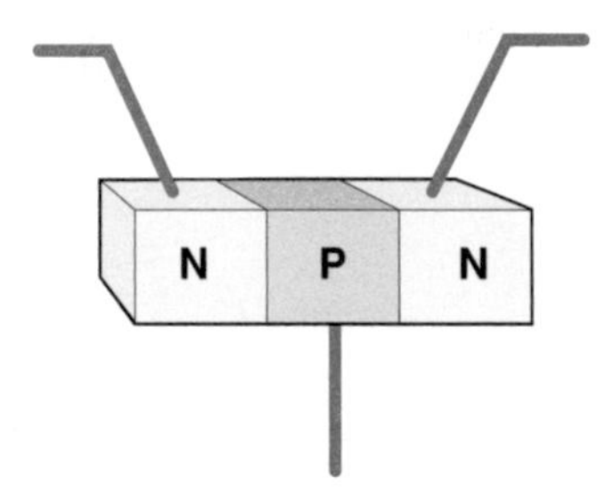

b. Silicon Configuration Suggested by the Symbol

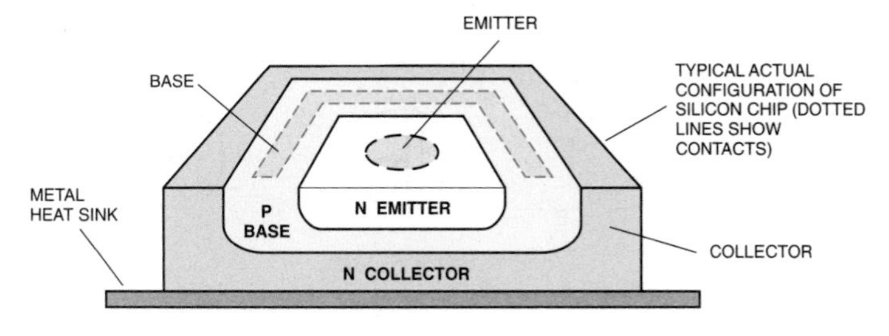

c. Diffused Sandwich Construction

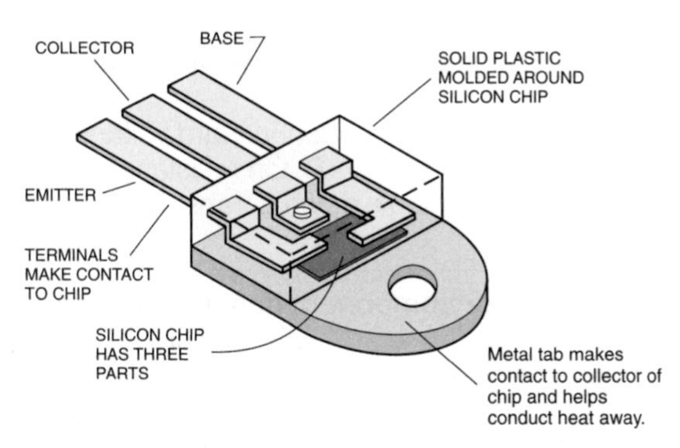

d. Transistor in Package

T6B05 Which of the following electronic components can amplify signals?

A. Transistor.
B. Variable resistor.
C. Electrolytic capacitor.
D. Multi-cell battery.

Here is that transistor once again! It can act as a switch, and can also amplify signals. Remember ***transistor***. **ANSWER A.**

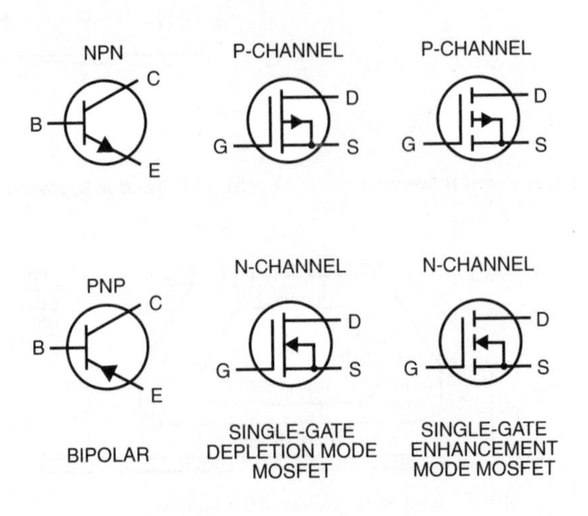

T6B12 What is the term that describes a transistor's ability to amplify a signal?

A. Gain.
B. Forward resistance.
C. Forward voltage drop.
D. On resistance.

We use the word ***gain*** to describe how much a transistor is ***able to amplify*** a weak signal. **ANSWER A.**

T6B10 Which semiconductor component has an emitter electrode?

A. Bipolar transistor.
B. Field effect transistor.
C. Silicon diode.
D. Bridge rectifier.

The common workhorse ***bipolar transistor*** comes in two varieties. NPN and PNP, depending on the three divisions of semiconductor materials layered within the transistor. B is for base, C is for the collector, and E is always represented by an arrow, the ***emitter***. If the arrow is pointing in, it is a PNP configuration. If the arrow is not pointing in, it is NPN. The transistors may be used as amplifiers, switches, oscillators, and are quite rugged and nearly immune to wearing out. **ANSWER A.**

T6B04 Which of these components is made of three layers of semiconductor material?

A. Alternator.
B. Bipolar junction transistor.
C. Triode.
D. Pentagrid converter.

The ***bipolar junction transistor has 3 layers of semiconductor materials***, with 2 junctions that form a PNP or NPN sandwich. These current-operated components might amplify, oscillate, and are the workhorses inside that brand new VHF/UHF handheld transceiver. **ANSWER B.**

T6B08 What does the abbreviation "FET" stand for?

A. Field Effect Transistor.
B. Fast Electron Transistor.
C. Free Electron Transition.
D. Field Emission Thickness.

Modern ham radio equipment use powerful transistors called field effect transistors, ***(FET)***. The ***Field Effect Transistor*** can accomplish many functions within our ham radio equipment, providing voltage amplification, where small changes in gate voltage can trigger large changes in current flow, leading to voltage amplification. **ANSWER A.**

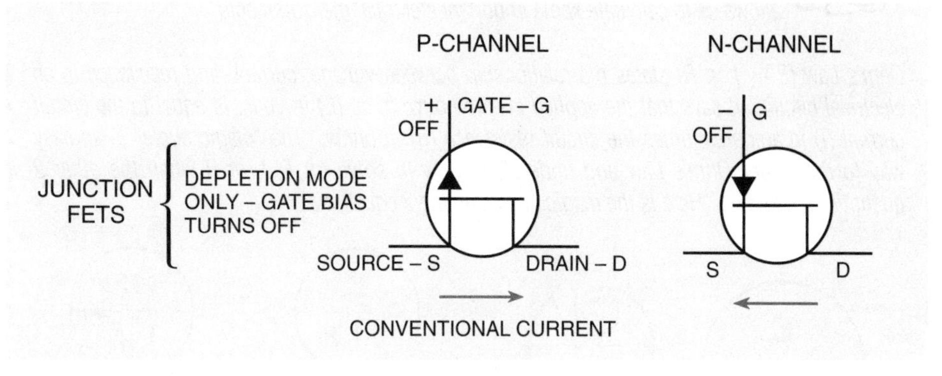

Junction FETs

T6B11 Which semiconductor component has a gate electrode?

A. Bipolar transistor.
B. Field effect transistor.
C. Silicon diode.
D. Bridge rectifier.

The ***Field Effect Transistor (FET)*** can be identified by its distinctive symbol, with the ***gate*** arrow pointing in or not pointing in. The voltage at the gate influences current through the transistor. **ANSWER B.**

Ohm's Law & the Magic Circle

Your Granddaddy Ham likely knew that a chap named Georg Simon Ohm (1789-1854) experimented with electricity and discovered that the resistance of a conductor depends on its length in feet, cross-sectional area in circular mils, and its resistivity, which is a parameter that depends on the molecular structure of the conductor and its temperature. Sounds complicated, and it is, but his discovery allows us to calculate some important electrical measurements.

Ohm's Law ($E = I \times R$) states the relationship between voltage, current, and resistance in an electrical circuit. It says that the applied electromotive force (E) in volts, is equal to the circuit current (I) in amperes, times the circuit resistance (R) in ohms. The "magic circle" is an easy way to remember Ohm's Law and understand how to solve for E, I, or R when the other 2 quantities are known. Here is the magic circle and the 3 equations:

$E = I \times R$
Finding Voltage

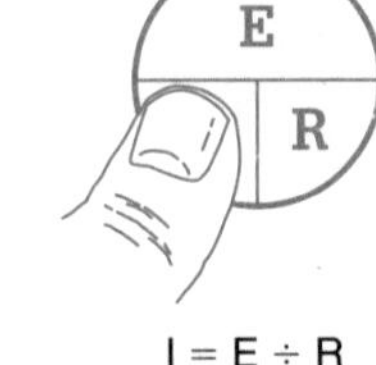

$I = E \div R$
Finding Current

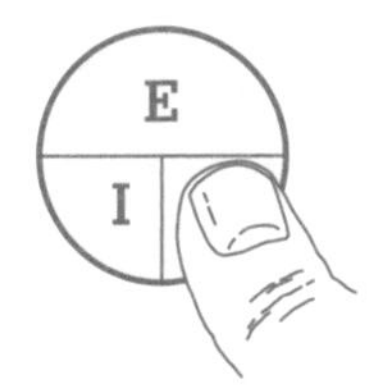

$R = E \div I$
Finding Resistance

To use the circle, cover the unknown quantity with your finger and solve the equation using the 2 known quantities. If you know the values of I and R and want to find the value of E, cover the E in the magic circle and it shows that you must multiple I times R. If you want to find I, cover the I and it shows that you must divided E by R. If you want to find R, cover the R and it shows you must divide E by I.

There is another "magic circle" to help you remember how to calculate power in a circuit. Power in watts (P) is equal to current (I) in amperes times volts (E) in volts. Use it the same way as the Ohm's Law magic circle; that is, cover the unknown quantity with your finger and perform the mathematical operation represented by the remaining quantities.

$P = I \times E$
Finding Power

$I = P \div E$
Finding Amperes

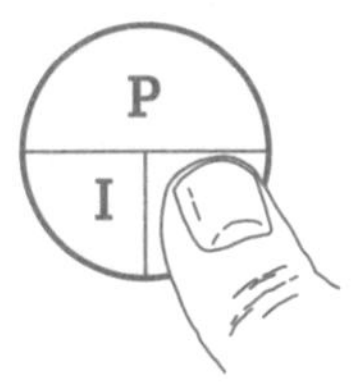

$E = P \div I$
Finding Voltage

It's The Law, Per Mr. OHM!

T5A10 Which term describes the rate at which electrical energy is used?

A. Resistance.
B. Current.
C. Power.
D. Voltage.

Go outside your home or condo and gaze at the ***power*** meter ***measuring the rate at which electrical energy is being consumed*** by all the gadgets inside your new ham shack! Luckily, your modern ham gear won't consume much energy so your power bill won't go up by more than a few cents per month! That power meter on the side of your house is called a watt meter, and it is calculating your household voltage times the amount of amperes of current flowing to keep your radio station powered up! **ANSWER C.**

T5A02 Electrical power is measured in which of the following units?

A. Volts.
B. Watts.
C. Ohms.
D. Amperes.

Power is energy, and power is measured in ***watts***. **ANSWER B.**

T5C08 What is the formula used to calculate electrical power in a DC circuit?

A. Power (P) equals voltage (E) multiplied by current (I).
B. Power (P) equals voltage (E) divided by current (I).
C. Power (P) equals voltage (E) minus current (I).
D. Power (P) equals voltage (E) plus current (I).

Power in watts is ***equal to volts times current*** in amps. A 100-watt light bulb, running on 110 VAC house voltage, will draw about 1 amp. The magic circle for power is: P over E I. Cover the unknown quantity with your finger, and perform the mathematical operation represented by the remaining quantities. **ANSWER A.**

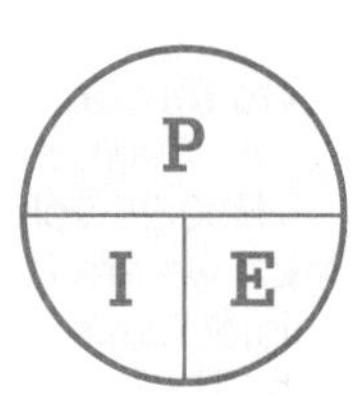

T5C09 How much power is being used in a circuit when the applied voltage is 13.8 volts DC and the current is 10 amperes?

A. 138 watts.
B. 0.7 watts.
C. 23.8 watts.
D. 3.8 watts.

Power is equal to volts times amps. In this problem, multiple 13.8 volts by 10 amps, and you end up with ***138 watts***. This one you can do in your head – easy as PIE!. **ANSWER A.**

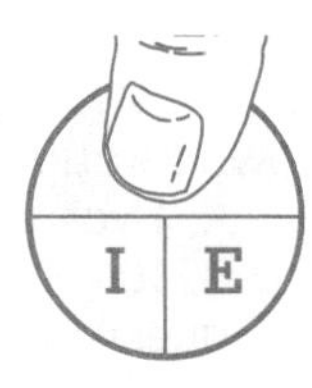

$P = I \times E$
Finding Power

T5C10 How much power is being used in a circuit when the applied voltage is 12 volts DC and the current is 2.5 amperes?

A. 4.8 watts.
B. 30 watts.
C. 14.5 watts.
D. 0.208 watts.

Power is equal to volts times amps. Multiple 12 volts by 2.5 amps, and you end up with ***30 watts***. **ANSWER B.**

T5C11 How many amperes are flowing in a circuit when the applied voltage is 12 volts DC and the load is 120 watts?

A. 0.1 amperes.
B. 10 amperes.
C. 12 amperes.
D. 132 amperes.

This time we are calculating for amps, so it is power (120) divided by voltage (12). Do the keystrokes: Clear Clear ***120 ÷ 12 = 10.*** Again, just because they list power second in the question, it still goes on top at P, and they list voltage first in the question, and that goes on the bottom as E. **ANSWER B.**

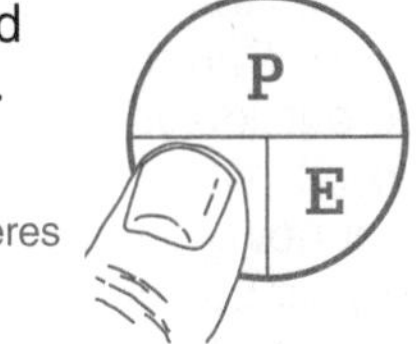

I = P ÷ E
Finding Amperes

T5D02 What formula is used to calculate voltage in a circuit?

A. Voltage (E) equals current (I) multiplied by resistance (R).
B. Voltage (E) equals current (I) divided by resistance (R).
C. Voltage (E) equals current (I) added to resistance (R).
D. Voltage (E) equals current (I) minus resistance (R).

Voltage = current × resistance, which is expressed as E = I × R. This simple formula, called Ohm's Law states the relationship between voltage, current and resistance in an electrical circuit. **ANSWER A.**

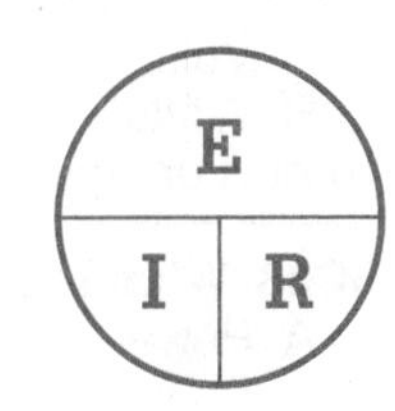

T5D10 What is the voltage across a 2-ohm resistor if a current of 0.5 amperes flows through it?

A. 1 volt.
B. 0.25 volts.
C. 2.5 volts.
D. 1.5 volts.

Since we are looking for E in this question, the voltage across a resistor, cover E with your finger, and you now have I (0.5 amps) times R (2 ohms). Simply multiple these 2 to obtain your answer of ***1 volt***. On your calculator, which is perfectly legal in the exam room, perform the following keystrokes: Clear Clear 0.5 × 2 = and the answer is 1 volt.. Commit the magic circle to memory now! **ANSWER A.**

E = I × R
Finding Voltage

T5D11 What is the voltage across a 10-ohm resistor if a current of 1 ampere flows through it?

A. 1 volt.
B. 10 volts.
C. 11 volts.
D. 9 volts.

The question starts out, "What is the voltage across…" so put your finger over E and see that the current in this question is 1 amp through a 10 ohm resistor. One multiplied by 10 is… ***10 volts***. This one you can do in your head. **ANSWER B.**

T5D12 What is the voltage across a 10-ohm resistor if a current of 2 amperes flows through it?

A. 8 volts.
B. 0.2 volts.
C. 12 volts.
D. 20 volts.

This question is looking for voltage, so we know it's going to be a simple multiplication of ***2 amperes*** through a ***10-ohm resistor***, with ***20 volts*** as the correct answer. **ANSWER D.**

T5D01 What formula is used to calculate current in a circuit?

A. Current (I) equals voltage (E) multiplied by resistance (R).
B. Current (I) equals voltage (E) divided by resistance (R).
C. Current (I) equals voltage (E) added to resistance (R).
D. Current (I) equals voltage (E) minus resistance (R).

For ***current***, put your finger over I, and it ***is voltage (E) divided by resistance (R)***. **ANSWER B.**

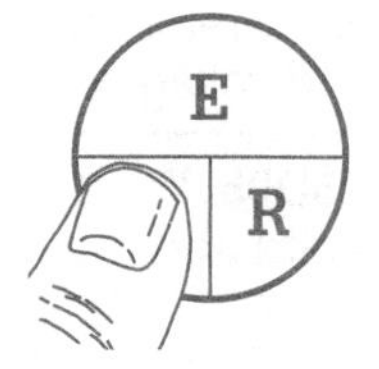

I = E ÷ R
Finding Current

T5D09 What is the current flowing through a 24-ohm resistor connected across 240 volts?

A. 24,000 amperes.
B. 0.1 amperes.
C. 10 amperes.
D. 216 amperes.

Do the keystrokes: Clear Clear ***240 ÷ 24 = 10***. Remember, to calculate current, it is voltage on top divided by resistance on the bottom. **ANSWER C.**

T5D08 What is the current flowing through a 100-ohm resistor connected across 200 volts?

A. 20,000 amperes.
B. 0.5 amperes.
C. 2 amperes.
D. 100 amperes.

Be careful on this question – they reversed the order of resistance and voltage which was in the previous question. In your magic circle, I = 200 ÷ 100. Calculator keystrokes: Clear Clear 200 (volts on the top) ÷ 100 ohms (on the bottom) = ***2 amperes***. **ANSWER C.**

T5D07 What is the current flow in a circuit with an applied voltage of 120 volts and a resistance of 80 ohms?

A. 9600 amperes.
B. 200 amperes.
C. 0.667 amperes.
D. 1.5 amperes.

Here they want to know current, so it is voltage (120 volts) divided by resistance (80 ohms). Here are your calculator keystrokes: Clear Clear 120 ÷ 80 = ***1.5***. Be careful that you don't reverse your division – they have an incorrect answer, C, just waiting for you! **ANSWER D.**

T5D03 What formula is used to calculate resistance in a circuit?

A. Resistance (R) equals voltage (E) multiplied by current (I).
B. Resistance (R) equals voltage (E) divided by current (I).
C. Resistance (R) equals voltage (E) added to current (I).
D. Resistance (R) equals voltage (E) minus current (I).

Put your finger over ***resistance*** in the magic circle, and see that it is ***voltage divided by current***. **ANSWER B.**

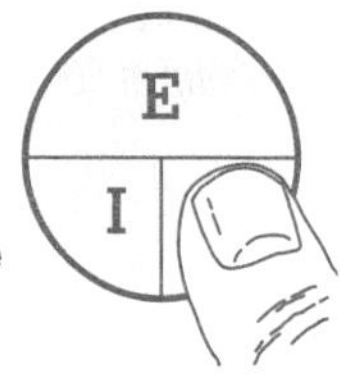

R = E ÷ I
Finding Resistance

T5D04 What is the resistance of a circuit in which a current of 3 amperes flows through a resistor connected to 90 volts?

A. 3 ohms. C. 93 ohms.
B. 30 ohms. D. 270 ohms.

Be careful – they list current first which would go in the bottom of your magic circle, and voltage at the top. Keystrokes: Clear Clear *90 ÷ 3 = 30.* **ANSWER B.**

T5D05 What is the resistance in a circuit for which the applied voltage is 12 volts and the current flow is 1.5 amperes?

A. 18 ohms. C. 8 ohms.
B. 0.125 ohms. D. 13.5 ohms.

In this problem, they list voltage first which is 12, on the top, divided by 1.5 amps on the bottom. Clear Clear *12 ÷ 1.5 = 8*. Read each question carefully because they switch around voltage and current, yet your magic circle always says put voltage on the top and current on the bottom when solving for resistance. **ANSWER C.**

T5D06 What is the resistance of a circuit that draws 4 amperes from a 12-volt source?

A. 3 ohms. C. 48 ohms.
B. 16 ohms. D. 8 Ohms.

Now remember, Gordo's rule – on most Technician Class questions, you divide the larger number by the smaller number, and presto, you end up with the correct answer. *12 divided by 4 equals 3, correct*? R = E (12 volts) ÷ I (4 amps). Ohm's Law – simple! **ANSWER A.**

Elmer Point: *Want to learn more about electricity and how electronics work? Here are two books I recommend highly for your self-education!*

Getting Started in Electronics *by Forrest M. Mims III is a true classic. It is used by a wide range of people – from junior-high teachers to the U.S. Army to teach the fundamentals of electricity and electronics.*

Basic Electronics *by Alvis J. Evans and Gene McWhorter goes a little deeper into the topic, and includes end of chapter quizzes and worked-out problems to teach you in detail the various aspects of electronics.*

Either book – or both – will give you a solid grounding in the theory, science, and practical applications of electronics.

You can get you copies of the books at your local ham radio store, on line at www.w5yi.org, or by calling The W5YI Group at 800-669-9594.

CD 4 TRACK 2

Picture This!

T6C01 What is the name for standardized representations of components in an electrical wiring diagram?

A. Electrical depictions.
B. Grey sketch.
C. Schematic symbols.
D. Component callouts.

As a new Technician Class operator, you should be familiar with the identification of simple schematic diagram illustrations. In the next few questions, we'll take a look at some schematic diagrams that you should know. We will home in on identifying ***schematic symbols*** in the ***schematic diagram*** for your exam. Don't panic! It's easy. **ANSWER C.**

T6C12 What do the symbols on a electrical circuit schematic diagram represent?

A. Electrical components.
B. Logic states.
C. Digital codes.
D. Traffic nodes.

Now, let's check ourselves. The ***symbols*** on an electrical circuit schematic diagram represent ***electrical components***. **ANSWER A.**

T6C13 Which of the following is accurately represented in electrical circuit schematic diagrams?

A. The wires lengths.
B. The physical appearance of components.
C. The way components are interconnected.
D. All of these choices are correct.

Schematic diagrams allow us to see exactly ***how components are interconnected***, right down to each and every lead. **ANSWER C.**

T6C10 What is component 3 in figure T3?

A. Connector.
B. Meter.
C. Variable capacitor.
D. Variable inductor.

Let's start with Figure 3. You probably will have only one of the three schematic figures on your exam. Component ***#3*** doesn't have a squiggly line like a resistor, but rather a coil type line, so it is a ***variable inductor***. It is variable because we see tap points on the hump lines and a line with an arrow indicating the inductor can be adjusted to any one of the taps. **ANSWER D.**

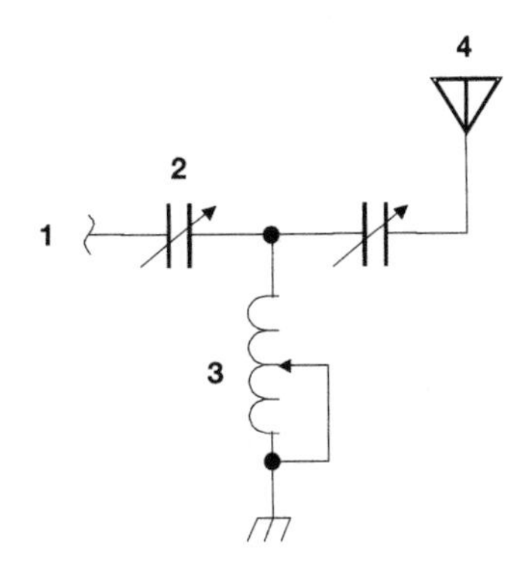

Figure T3

T6C11 What is component 4 in figure T3?

A. Antenna.
B. Transmitter.
C. Dummy load.
D. Ground.

Component ***#4*** looks like what it represents – an ***antenna***! That antenna is tuned by some of the preceding circuit components. **ANSWER A.**

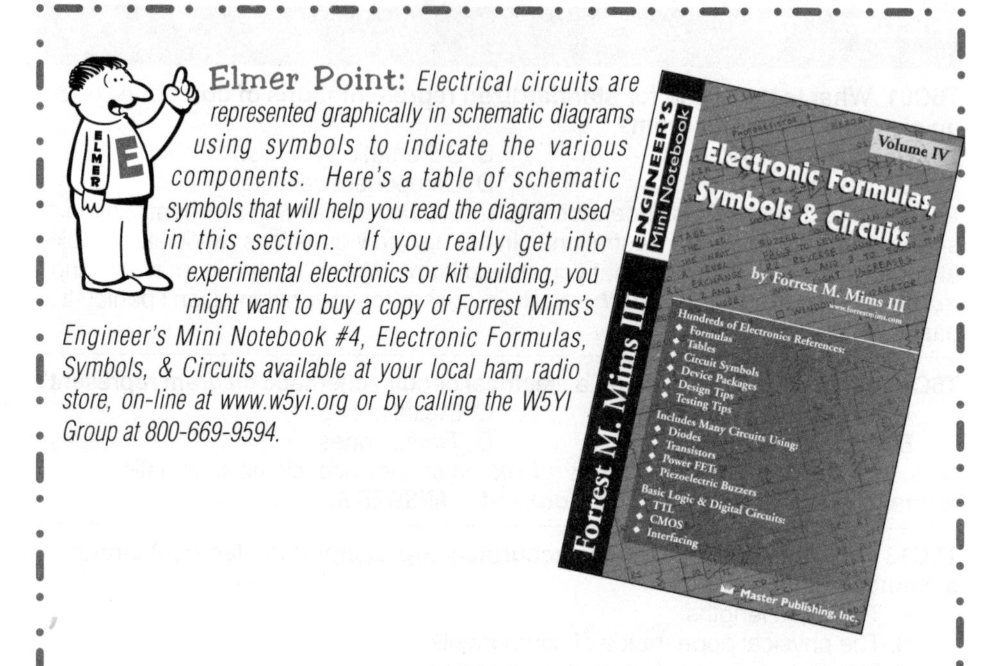

Elmer Point: *Electrical circuits are represented graphically in schematic diagrams using symbols to indicate the various components. Here's a table of schematic symbols that will help you read the diagram used in this section. If you really get into experimental electronics or kit building, you might want to buy a copy of Forrest Mims's Engineer's Mini Notebook #4, Electronic Formulas, Symbols, & Circuits available at your local ham radio store, on-line at www.w5yi.org or by calling the W5YI Group at 800-669-9594.*

SCHEMATIC SYMBOLS

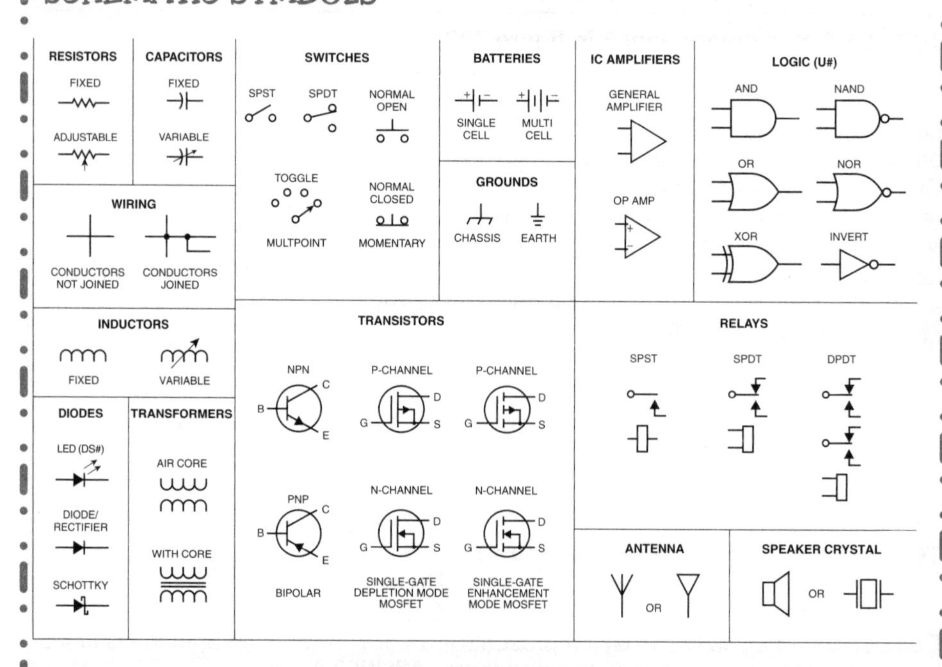

T6D08 Which of the following is used together with an inductor to make a tuned circuit?

A. Resistor.
B. Zener diode.
C. Potentiometer.
D. Capacitor.

We use series and parallel coils and capacitors to develop a tuned circuit inside your new radio. Another name for a coil is an inductor, and when used with a ***capacitor*** you now have a nice tuned circuit. Do you see the variable capacitor at #2 in figure T3? **ANSWER D.**

T6C02 What is component 1 in figure T1?

A. Resistor.
B. Transistor.
C. Battery.
D. Connector.

This figure will be found on your actual Technician Class exam sheet – either with the question or on the last page of your examination materials. Component #1, that ***squiggly line, is a resistor***. Can you imagine current flowing through all the squiggles offering resistance? **ANSWER A**

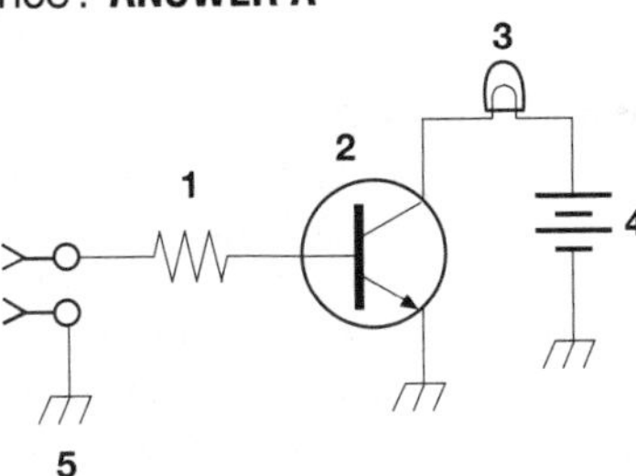

Figure T1

T6C03 What is component 2 in figure T1?

A. Resistor.
B. Transistor.
C. Indicator lamp.
D. Connector.

Component ***#2*** is our friendly ***transistor***. The arrow is NOT pointing in, so it is an NPN transistor. **ANSWER B.**

T6D10 What is the function of component 2 in Figure T1?

A. Give off light when current flows through it.
B. Supply electrical energy.
C. Control the flow of current.
D. Convert electrical energy into radio waves .

Component ***#2*** is a NPN ***transistor***, and this one can ***control the flow of current***, much like a valve. **ANSWER C.**

T6C04 What is component 3 in figure T1?

A. Resistor.
B. Transistor.
C. Lamp.
D. Ground symbol.

Component ***#3*** looks just like it is – a small indicator lamp. ***Looks like a lamp***, doesn't it? **ANSWER C.**

T6C05 What is component 4 in figure T1?

A. Resistor.
B. Transistor.
C. Battery.
D. Ground symbol.

It takes voltage to make this circuit work, and we get the voltage from the battery, component ***#4***. **ANSWER C.**

T6D03 What type of switch is represented by item 3 in figure T2?

A. Single-pole single-throw. C. Double-pole single-throw.
B. Single-pole double-throw. D. Double-pole double-throw.

Your handheld may have one of these – a ***single-pole, single-throw switch***. It is single in both senses because you see only one wire going to the switch, and only one single contact point. Single pole-single throw. **ANSWER A.**

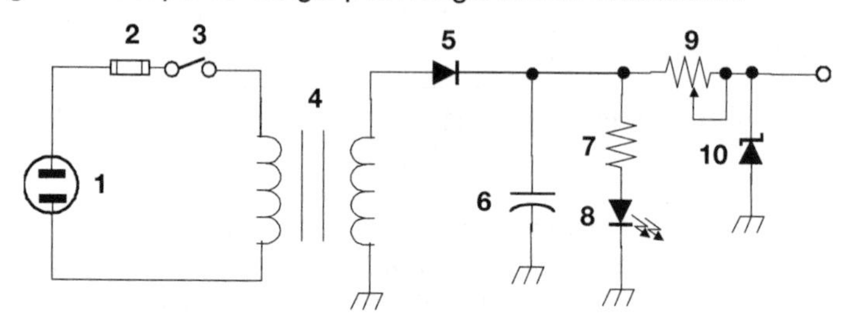

Figure T2

T6C09 What is component 4 in figure T2?

A. Variable inductor. C. Potentiometer.
B. Double-pole switch. D. Transformer.

Component *#4* takes in everything around it, and is a transformer. Voltage is passed from the windings on the left, to the windings on the right, with the two vertical lines representing an iron core. This ***transformer*** looks to have about the same number of turns on the primary and secondary, so the voltage going in will be about the same amount of voltage coming out the other side! **ANSWER D.**

T6C06 What is component 6 in figure T2?

A. Resistor. C. Regulator IC.
B. Capacitor. D. Transistor.

Ok, you're doing so well, let's turn you into a master engineer, and look at Figure T2. Component *#6* has 2 parallel (sort of) plates, separated by an insulation, so it must be a ***capacitor***. **ANSWER B.**

T6C07 What is component 8 in figure T2?

A. Resistor. C. Regulator IC.
B. Inductor. D. Light emitting diode.

Component #8 gives it away – see the little arrow symbols showing the effects of light? Component *#8* is a ***light emitting diode*** — an ***LED***. Easy! **ANSWER D.**

T6C08 What is component 9 in figure T2?

A. Variable capacitor. C. Variable resistor.
B. Variable inductor. D. Variable transformer.

Component *#9* is indeed a resistor, but it has a variable tap point on it, so it is simply a ***variable resistor***. We'll formally call it a potentiometer and this could be the volume control on your handheld. **ANSWER C.**

T6D04 Which of the following can be used to display signal strength on a numeric scale?

A. Potentiometer. C. Meter.
B. Transistor. D. Relay.

Your larger, high-frequency transceivers may have a mechanical ***meter*** movement to illustrate incoming ***signal strength***. Even if it is an LED or LCD readout, we still call it a signal strength meter. **ANSWER C.**

T6D02 What best describes a relay?

A. A switch controlled by an electromagnet.
B. A current controlled amplifier.
C. An optical sensor.
D. A pass transistor.

Most of your handhelds don't have one, but a mobile radio that puts out 50 watts will likely contain a relay. The ***relay is a mechanical switch***, opened and closed by current passing through a coil, creating an ***electromagnet***. As soon as the coil is energized, the switch goes from one state to another. **ANSWER A.**

T5B09 What is the approximate amount of change, measured in decibels (dB), of a power increase from 5 watts to 10 watts?

A. 2 dB.
B. 3 dB.
C. 5 dB.
D. 10 dB.

Doubling your power will lead to a power increase of 3 dB. Halving your power will lead to a 3 dB decrease. But since we are going ***from 5 watts to 10 watts***, we are doubling our power, and that is ***a 3 dB increase***. **ANSWER B.**

Decibels

It is important to know about decibels (dB) because they are used extensively in electronics. The decibel is used to describe a change in power levels. It is a measure of the ratio of power output (P_1) to power input (P_2). It is relitavely simple to calculate a dB power change. Remember this table:

dB	Power Change
3 dB	2× Power change
6 dB	4× Power change
9 dB	8× Power change
10 dB	10× Power change
20 dB	100× Power change
30 dB	1000× Power change
40 dB	10,000× Power change

Derivation:

If $dB = 10 \log_{10} \frac{P_1}{P_2}$

then what power ratio is 20 dB?

$$20 = 10 \log_{10} \frac{P_1}{P_2}$$

$$\frac{20}{10} = \log_{10} \frac{P_1}{P_2}$$

$$2 = \log_{10} \frac{P_1}{P_2}$$

Remember: logarithm of a number is the exponent to which the base must be raised to get the number.

$$\therefore 10^2 = \frac{P_1}{P_2}$$

$$100 = \frac{P_1}{P_2}$$

Or $P_1 = 100\, P_2$

20 dB means P_1 is 100 times P_2

T5B10 What is the approximate amount of change, measured in decibels (dB), of a power decrease from 12 watts to 3 watts?

A. 1 dB.
B. 3 dB.
C. 6 dB.
D. 9 dB.

Now we go ***from 12 watts down to 3 watts*** by pushing the low power button on our small mobile radio. This is a 4 times DECREASE. ($4 \times 3 = 12$) A 4 times decrease is a power ***decrease of 6 dB***. **ANSWER C.**

T5B11 What is the approximate amount of change, measured in decibels (dB), of a power increase from 20 watts to 200 watts?

A. 10 dB.
B. 12 dB.
C. 18 dB.
D. 28 dB.

Now we add a linear amplifier. Normally, I don't recommend any linear amplifiers as you are getting started as a Technician Class operator. Going ***from 20 watts to 200 watts*** is a bit dangerous, and that is a 10 times increase in power ($20 \times 10 = 200$). Ten times increase ***equals 10 dB***. **ANSWER A.**

T6D05 What type of circuit controls the amount of voltage from a power supply?

A. Regulator.
B. Oscillator.
C. Filter.
D. Phase inverter.

If you want to run your handheld on house power, you'll need a power supply that will provide regulated 12 volts to your handheld input circuit. This power supply must have a good ***regulator*** built in so that it does not exceed the 12 volts DC input that your handheld works with. Never transmit from your handheld when plugged in to an AC "wall wart" because there is not enough filtering within that "wall wart" for a good signal. It is okay for listening, but unplug before transmitting! **ANSWER A.**

T6D06 What component is commonly used to change 120V AC house current to a lower AC voltage for other uses?

A. Variable capacitor.
B. Transformer.
C. Transistor.
D. Diode.

When you purchase your new dual-band handheld it will come with a wall charger that plugs into the side of your radio. The wall charger contains a small ***transformer*** that takes 120 volts AC on the primary and steps it down to a lower AC voltage on the secondary. Diodes and capacitors then filter this lowered AC and convert it to 12 volts DC. The common "wall wart" contains all of this circuitry, and newer "switcher" wall warts are much lighter in weight and the transformer is extremely small sized! **ANSWER B.**

T6D09 What is the name of a device that combines several semiconductors and other components into one package?

A. Transducer.
B. Multi-pole relay.
C. Integrated circuit.
D. Transformer.

If you ever look inside the modern ham radio, you will see rectangular "chips" that are large scale ***integrated circuits***. These chips contain thousands of semiconductors in one nice neat package – abbreviated ***"ICs"***. **ANSWER C.**

Large-scale integrated circuit chips in a PLL section of a communications receiver.

T6B07 What does the abbreviation "LED" stand for?

A. Low Emission Diode.
B. Light Emitting Diode.
C. Liquid Emission Detector.
D. Long Echo Delay.

LED stands for Light Emitting Diode, and this is that green, red, or amber indicator that comes up on your handheld when you are transmitting. The LED draws almost no amount of current and will last for hundreds of thousands of hours without burnout. This is a big improvement over our older radios with their tiny, grain-of-wheat light bulbs. ***LED = Light Emitting Diode***. **ANSWER B.**

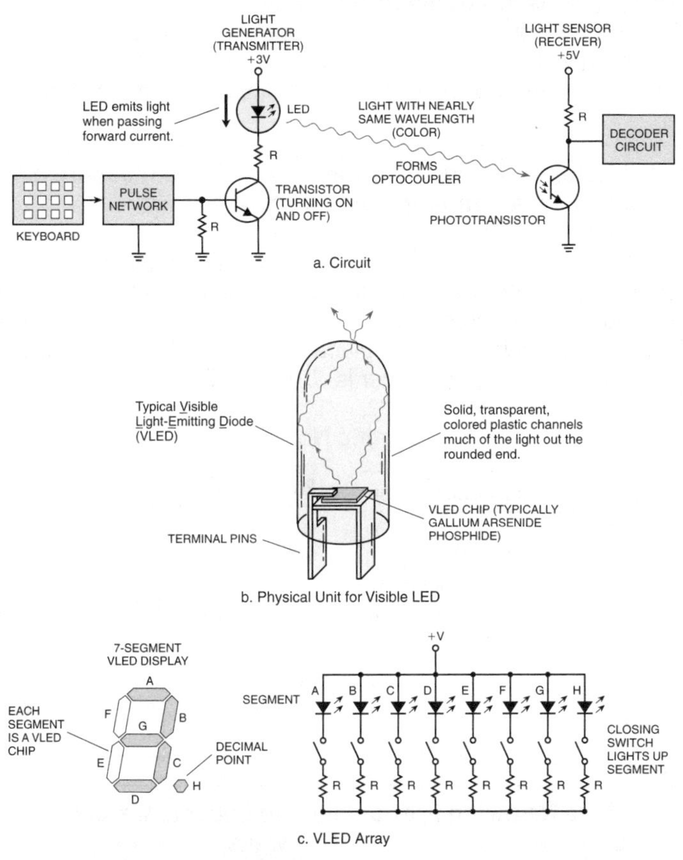

An array of LEDs and resistors mounted on a printed circuit board.

T6D07 Which of the following is commonly used as a visual indicator?

A. LED.
B. FET.
C. Zener diode.
D. Bipolar transistor.

A good ***visual indicator*** on a handheld radio is the ***LED*** — the light emitting diode that is often used as a transmit indicator. **ANSWER A.**

T5B02 What is another way to specify a radio signal frequency of 1,500,000 hertz?

A. 1500 kHz.
B. 1500 MHz.
C. 15 GHz.
D. 150 kHz.

To keep you from running out of pencil lead, we can abbreviate ***1,500,000 Hz*** as either ***1,500 kHz***, or 1.5 MHz. From Hz to kHz, move the decimal 3 places to the left. From kHz to MHz, move it 3 more places to the left. **ANSWER A.**

T5B03 How many volts are equal to one kilovolt?

A. One one-thousandth of a volt.
B. One hundred volts.
C. One thousand volts.
D. One million volts.

Remember kilo? ***Kilo means one thousand***, Answer C. Watch out for Answer A. **ANSWER C.**

T5B06 If an ammeter calibrated in amperes is used to measure a 3000-milliampere current, what reading would it show?

A. 0.003 amperes.
B. 0.3 amperes.
C. 3 amperes.
D. 3,000,000 amperes.

One milliampere equals one one-thousandth of an ampere (1×10^{-3}); therefore, one ampere equal 1000 milliamperes. Divide milliamperes by 1000 to convert to amperes. Or move the decimal point 3 places to the left. Calculator keystrokes are: CLEAR 3000 ÷ 1000 = and the answer is ***3***. **ANSWER C.**

Scientific Notation

Prefix	Symbol			Multiplication Factor
exa	E	10^{18}	=	1,000,000,000,000,000,000
peta	P	10^{15}	=	1,000,000,000,000,000
tera	T	10^{12}	=	1,000,000,000,000
giga	G	10^{9}	=	1,000,000,000
mega	M	10^{6}	=	1,000,000
kilo	k	10^{3}	=	1,000
hecto	h	10^{2}	=	100
deca	da	10^{1}	=	10
(unit)		10^{0}	=	1

Prefix	Symbol			Multiplication Factor
deci	d	10^{-1}	=	0.1
centi	c	10^{-2}	=	0.01
milli	m	10^{-3}	=	0.001
micro	μ	10^{-6}	=	0.000001
nano	n	10^{-9}	=	0.000000001
pico	p	10^{-12}	=	0.000000000001
femto	f	10^{-15}	=	0.000000000000001
atto	a	10^{-18}	=	0.000000000000000001

T5B05 Which of the following is equivalent to 500 milliwatts?

A. 0.02 watts.
B. 0.5 watts.
C. 5 watts.
D. 50 watts.

Your handheld transceiver can be dialed down to minimum power output, dramatically conserving battery life. Five hundred milliwatts can be converted to watts by moving the decimal point 3 places to the left. So a handheld at 500 milliwatts output is transmitting ***0.5 watts*** of power, which is the same as a half watt of power. Believe it or not, you can make many contacts through local repeaters at a half-watt of power, and your batteries will love you for it. **ANSWER B.**

T5B01 How many milliamperes is 1.5 amperes?

A. 15 milliamperes.
B. 150 milliamperes.
C. 1,500 milliamperes.
D. 15,000 milliamperes.

Your new dual-band handheld might offer as much as 5 to 7 watts of output power. Depending on the battery pack voltage, the transmitter could actually draw as much as 1,500 milliamperes. But don't panic with all those numbers – move the decimal point 3 places to the left to go from milliamps to amps. In this question, to convert 1.5 amps to ***1500 milliamperes***, move the decimal point 3 places to the right. Three places. **ANSWER C.**

T5B08 How many microfarads are 1,000,000 picofarads?

A. 0.001 microfarads.
B. 1 microfarad.
C. 1000 microfarads.
D. 1,000,000,000 microfarads.

A picofarad is one millionth (1×10^{-6}) of a microfarad. Move the decimal point 6 places to the left to convert picofarads to microfarads. ***1 million pico equals 1 microfarad***. **ANSWER B.**

T5B04 How many volts are equal to one microvolt?

A. One one-millionth of a volt.
B. One million volts.
C. One thousand kilovolts.
D. One one-thousandth of a volt.

We measure the receiver capabilities on a handheld in microvolts. The word ***"micro" means one-millionth***. Mega means million, kilo means thousand, and milli means thousandth. Micro means one one-millionth. **ANSWER A.**

T7D08 Which of the following types of solder is best for radio and electronic use?

A. Acid-core solder.
B. Silver solder.
C. Rosin-core solder.
D. Aluminum solder.

Rosin-core solder is commonly available at the same place you purchase your soldering iron or soldering gun. A little soldering pen is fine for working on tiny circuits, but you're going to need a massive soldering iron or a big 200 watt soldering gun if you plan to install coax connectors properly. Always use ***rosin-core solder***. Wear protective glasses, too. **ANSWER C.**

T7D09 What is the characteristic appearance of a "cold" solder joint?

A. Dark black spots.
B. A bright or shiny surface.
C. A grainy or dull surface.
D. A greenish tint.

It's easy to tell if you've made a good solder connection – the solder looks shiny. However, a ***"cold,"*** poorly-***soldered joint looks grainy and dull***. **ANSWER C.**

T7D07 Which of the following measurements are commonly made using a multimeter?

A. SWR and RF power.
B. Signal strength and noise.
C. Impedance and reactance.
D. Voltage and resistance.

Every amateur operator should own a ***multimeter***. The multiple function meter can measure ***voltage, current, and resistance***, and check continuity. Even an inexpensive multimeter is better than no meter when you are trying to check out a circuit in the field. You can buy an excellent multimeter for less than $25.00. **ANSWER D.**

Parameter	Basic Unit	Measuring Instrument
Voltage (E)	Volts	Voltmeter
Current (I)	Amperes	Ammeter
Resistance (R)	Ohms	Ohmmeter
Power (P)	Watts	Wattmeter

T7D11 Which of the following precautions should be taken when measuring circuit resistance with an ohmmeter?

A. Ensure that the applied voltages are correct.
B. Ensure that the circuit is not powered.
C. Ensure that the circuit is grounded.
D. Ensure that the circuit is operating at the correct frequency.

Any time you are checking a circuit with an ***Ohm meter***, make sure the ***circuit is NOT energized***. If you check any circuit with voltage on it, you will probably toast the Ohm meter for life. **ANSWER B.**

T7D06 Which of the following might damage a multimeter?

A. Measuring a voltage too small for the chosen scale.
B. Leaving the meter in the milliamps position overnight.
C. Attempting to measure voltage when using the resistance setting.
D. Not allowing it to warm up properly.

You're likely to ***damage*** your brand new needle ***multimeter by measuring voltage*** if you accidentally leave it ***in the ohms reading setting***. **ANSWER C.**

T7D10 What is probably happening when an ohmmeter, connected across a circuit, initially indicates a low resistance and then shows increasing resistance with time?

A. The ohmmeter is defective.
B. The circuit contains a large capacitor.
C. The circuit contains a large inductor.
D. The circuit is a relaxation oscillator.

If you have a big old capacitor hanging around the shack, keep it handy! That big old capacitor has a telltale signature when you check it with an Ohm meter. On a high Ohms reading scale, the ***discharged capacitor will*** first look like an almost short circuit, and then ***show increasing resistance*** as the capacitor begins to charge up from your Ohm meter test. OK, that's nice, but how can we use this principle in the real world of ham radio? Let's say you brought in 5 lines of coaxial cable from roof to the shack and you forgot to tag which one goes where. Use a couple of alligator clip cables to take the big electrolytic capacitor, and put it across one coaxial cable end. Now head for the roof. Put your Ohm meter on an intermediate scale, and start testing each cable. All but one will look like an open connection, other than the one terminated with the capacitor, which first looks like low resistance, and then you see the needle meter show an increase in resistance with time. You're so smart! **ANSWER B.**

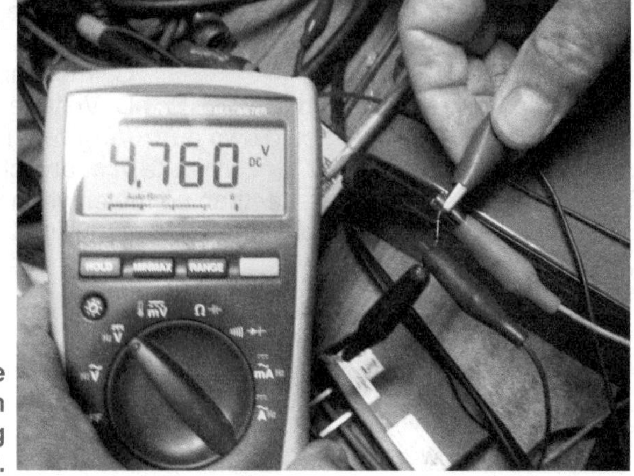

Learning how to use a multimeter is an essential skill in testing and repairing radio gear.

Antennas

T9A03 Which of the following describes a simple dipole mounted so the conductor is parallel to the Earth's surface?

A. A ground wave antenna.
B. A horizontally polarized antenna.
C. A rhombic antenna.
D. A vertically polarized antenna.

Plenty of new Technician Class operators, looking to work some skywaves on 6- and 10-meters, create their own halfwave dipole antennas. ***The dipole antenna*** is usually mounted parallel to the Earth's surface to transmit ***horizontal*** waves. **ANSWER B.**

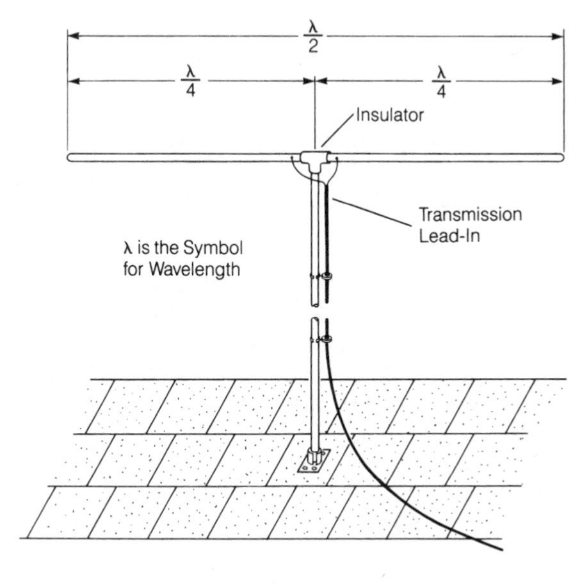

Dipole Antenna

T9A10 In which direction is the radiation strongest from a half-wave dipole antenna in free space?

A. Equally in all directions.
B. Off the ends of the antenna.
C. Broadside to the antenna.
D. In the direction of the feed line.

If the dipole is erected east and west, the energy would go out mostly north and south, ***broadside to the antenna***. Slightly droop the dipole ends if you wish more energy in other directions. **ANSWER C.**

T9A09 What is the approximate length, in inches, of a 6 meter 1/2-wavelength wire dipole antenna?

A. 6.
B. 50.
C. 112.
D. 236.

To calculate the length, end to end, of a ***6-meter, halfwave, wire dipole*** antenna, we first need to convert 6 meters over to megahertz. Remember that formula, 300 divided by the frequency in megahertz to equal meters? Three hundred divided by 6 equals approximately 50 MHz, and now apply the formula for a halfwave dipole: 468 ÷ frequency in MHz = a halfwave dipole in feet. Calculator: 468 ÷ 50 = 9.36 feet. Now, multiply 9.36 × 12 = to convert feet to inches, and you get ***112.3 inches***. You're a regular whiz on that keypad! **ANSWER C.**

T9A05 How would you change a dipole antenna to make it resonant on a higher frequency?

A. Lengthen it.
B. Insert coils in series with radiating wires.
C. Shorten it.
D. Add capacity hats to the ends of the radiating wires.

You can approximate what band a ham is transmitting on by looking at their dipole. The dipole is usually one-half the wavelength length of the meter band of operation. For instance, I can spot a 6-meter, brand new, Technician Class dipole as approximately 3 meters long, end to end, with coax feeding to it in the center. To ***increase*** the antenna's ***resonant frequency, shorten*** it by twisting back a couple of inches of wire at each end and it will work at a slightly higher frequency. **ANSWER C.**

T9A02 Which of the following is true regarding vertical antennas?

A. The magnetic field is perpendicular to the Earth.
B. The electric field is perpendicular to the Earth.
C. The phase is inverted.
D. The phase is reversed.

If an antenna element is perpendicular to the Earth's surface, it is standing straight up and down – it is a ***vertical antenna*** and will radiate vertical electric field radio waves. To get started, consider a dual-band 2 meter/440 MHz vertical home antenna, available for under $100. Just add coax cable between the vent-mounted antenna and your transceiver, and you are on the air. No tuning required for your little handheld or mobile 2 band radio. These antennas are usually white fiberglass with internal 2- band copper elements. Some are one section, some two section, and you can get a tall three-section dual-band antenna, too, if you live in an extremely remote area and need the most gain. Usually, the one- or two-section antenna, standing less than 10 feet tall, will do the trick nicely. Get the antenna up on your roof, be careful in how you get the antenna placed, watch for wires, and enjoy great range on 2 meters and 440 MHz. **ANSWER B.**

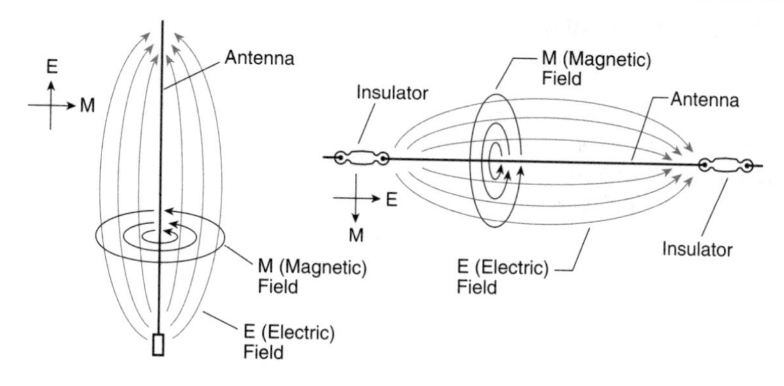

Horizontal and Vertical Polarization

T9A08 What is the approximate length, in inches, of a quarter-wavelength vertical antenna for 146 MHz?

A. 112.
B. 50.
C. 19.
D. 12.

We calculate the length, in feet, of a ***quarter-wavelength*** vertical antenna by ***dividing 234 by the antenna's operating frequency*** in megahertz. Let's try the calculator on this one: Clear Clear ***234 ÷ 146 = 1.6 feet***. Since they want the answer in inches, no big deal... multiply ***1.6 × 12 = 19 inches approximately***. To convert feet and fractions of a foot to inches, multiple by 12. **ANSWER C.**

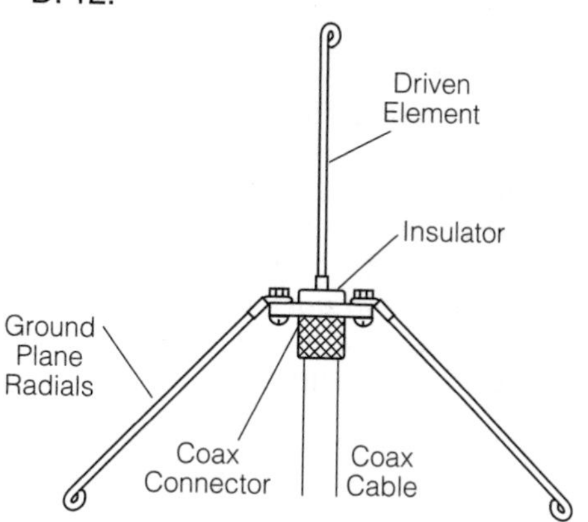

Vertical 1/4 λ Ground-Plane Antenna

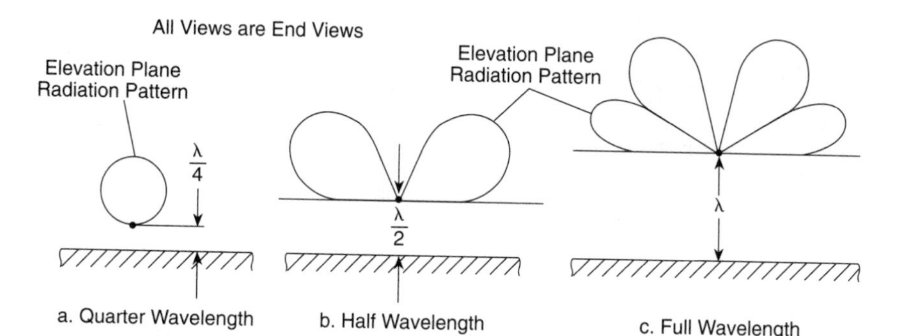

The Radiation Pattern of an Antenna Changes as Height Above Ground is Varied
Source: *Antennas,* A.J. Evans, K.E. Britain, ©1998, Master Publishing, Niles, Illinois

T9A06 What type of antennas are the quad, Yagi, and dish?

A. Non-resonant antennas.
B. Loop antennas.
C. Directional antennas.
D. Isotropic antennas.

Once you are on the air with your dual-band equipment, you may wish to reach out a little further in one specific direction. Antennas like the ***quad, the Yagi, and the microwave dish are all highly directional*** – they beam your signal in one specific direction. **ANSWER C.**

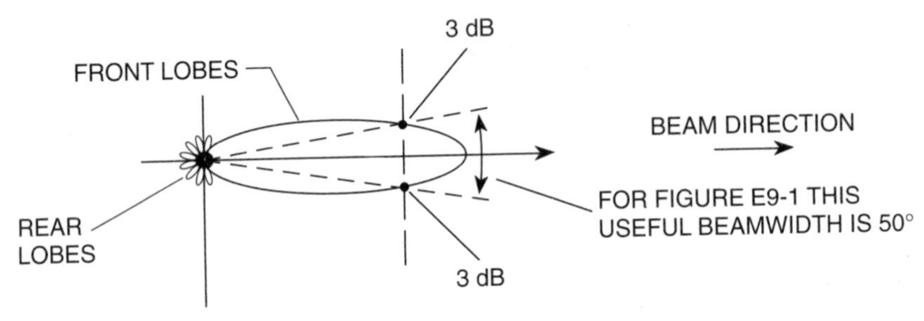

Directional Radiation Pattern of a Yagi Beam

Source: *Antennas – Selection and Installation*, A.J. Evans, Copyright ©1986 Master Publishing, Inc., Niles, Illinois

T9A01 What is a beam antenna?

A. An antenna built from aluminum I-beams.
B. An omnidirectional antenna invented by Clarence Beam.
C. An antenna that concentrates signals in one direction.
D. An antenna that reverses the phase of received signals.

The beam antenna is much like that new digital over-the-air television antenna on the roof – elements all in line with each other, with the shorter elements in the front. ***This antenna will concentrate signals in one direction***. The beam antenna is excellent for satellite and weak-signal work, but not all that necessary for talking around town on 2 meters and 70 centimeters, the 440 MHz band. **ANSWER C.**

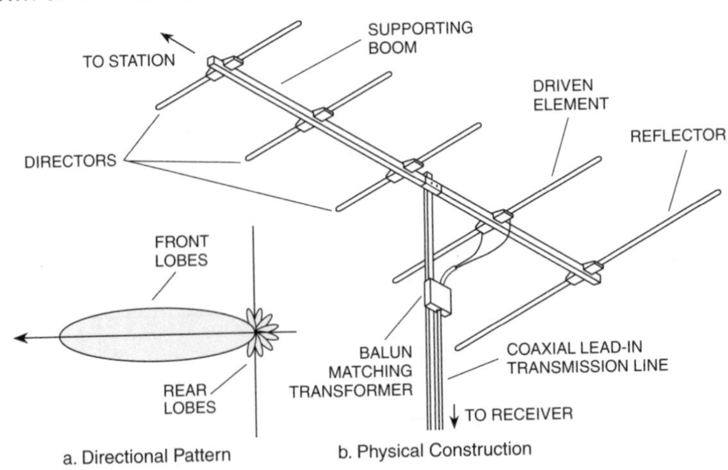

A Beam Antenna — The Yagi Antenna

Source: *Antennas – Selection and Installation,* ©1986, Master Publishing, Inc., Niles, Illinois

T8C01 Which of the following methods is used to locate sources of noise interference or jamming?

A. Echolocation.
B. Doppler radar .
C. Radio direction finding.
D. Phase locking.

You just added some home office equipment and notice there is a steady, no-voice carrier sitting right on your favorite repeater channel. You know it's coming from your house because your handheld out on the street is just fine. Practice your ***radio direction finding*** skills by going into your home office, tuning in the steady "dead carrier," and start switching off different pieces of equipment and pulling the plug from the power cord. Remember, some pieces of office equipment will still continue to transmit phantom signals on ham bands, although turned off, yet still plugged in. When you pull the plug that drops the signal, time to replace that particular piece of equipment with another brand that will hopefully not cause interference. Worst offenders in the home office? FAX machines, printers, and older telephone modems. **ANSWER C.**

W6JAY practices his fox hunt direction-finding skills

T8C02 Which of these items would be useful for a hidden transmitter hunt?

A. Calibrated SWR meter.
B. A directional antenna.
C. A calibrated noise bridge.
D. All of these choices are correct.

A very exciting sport is called "fox hunting." A ham will hide a low-power transmitter on the 2-meter band, usually transmitting around 146.565. Expert transmit hunters will use a ***directional antenna*** to beam into the general signal direction. The closer they get to the transmitter the stronger the signal appears on their handheld. When they get relatively close, they pull off the directional antenna, and then start "sniffing" around with their handheld with maybe just a paperclip stuck in the antenna socket. **ANSWER B.**

Gordo and Chip, K7JA, fox-hunting with a 2-meter cubical quad antenna.

Ham Hint: *Transmitter hunting is not only a fun sport, but a great skill to work with rescue groups tracking down personal locator beacons and ELTs activated in the mountains, or an EPIRB activated out on the ocean. Transmitter hunting is an international sport, and you better be in good shape to do a lot of running when the competition is right behind you homing in on that tiny hidden signal! WARNING: Any time you are around a lot of other transmitter hunters, be sure to wear protective lenses because everyone is swinging those beam antennas all over the place, and you want to be Mr. Safety.*

T3A05 When using a directional antenna, how might your station be able to access a distant repeater if buildings or obstructions are blocking the direct line of sight path?

A. Change from vertical to horizontal polarization.
B. Try to find a path that reflects signals to the repeater.
C. Try the long path.
D. Increase the antenna SWR.

A ***directional beam antenna*** for home use may bounce a signal to and from a distant repeater off of a nearby building or metal billboard! **ANSWER B.**

T9A11 What is meant by the gain of an antenna?

A. The additional power that is added to the transmitter power.
B. The additional power that is lost in the antenna when transmitting on a higher frequency.
C. The increase in signal strength in a specified direction when compared to a reference antenna.
D. The increase in impedance on receive or transmit compared to a reference antenna.

The term "***gain of an antenna***" refers to the ***increase in signal strength in a specific direction*** from an antenna when it is compared to the radiation pattern of an isotropic radiator (the reference antenna). The isotropic radiator is an imaginary antenna that is a sphere that radiates signals equally well in all directions. An antenna with GAIN is one that takes energy from somewhere within the antenna pattern, and combines it with energy in the desired direction of transmission. Got it? It's like taking a reflector to a plain old light bulb, and taking wasted energy going up, and redirecting it with combined energy down at street level. Your mobile VHF and UHF 2 meter and 440 MHz antennas all have a little bit of omnidirectional gain, taking energy that would normally go straight up, and reradiating it down close to the horizon. **ANSWER C.**

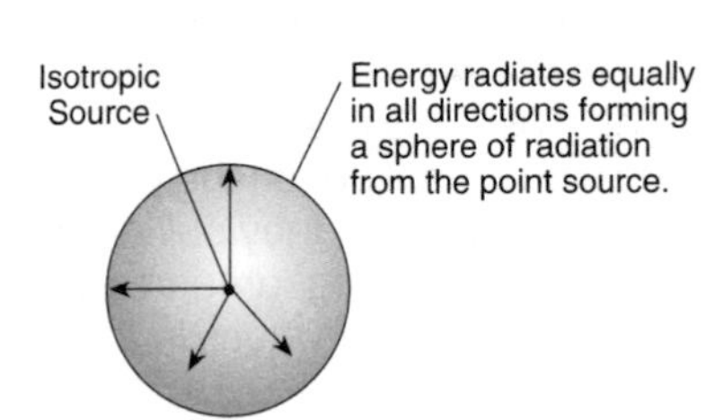

Isotropic Radiator Pattern

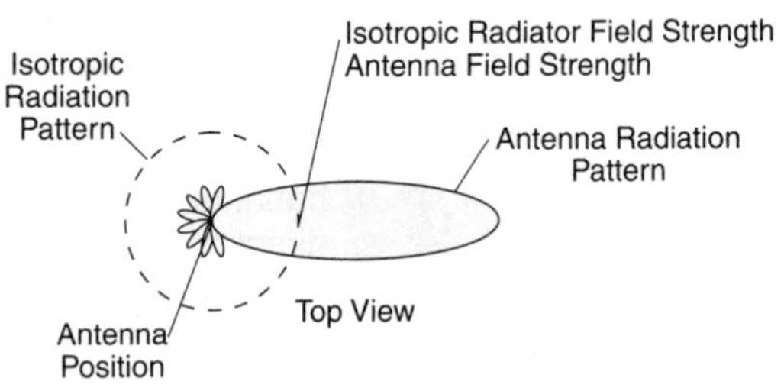

"Gain" of an antenna.

T3A03 What antenna polarization is normally used for long-distance weak-signal CW and SSB contacts using the VHF and UHF bands?

A. Right-hand circular.
B. Left-hand circular.
C. Horizontal.
D. Vertical.

As a new Technician Class ham radio operator, your first radio will likely be a dual band handheld, operating FM (frequency modulation), with a vertical antenna. If you enjoy talking with a lot of other hams, about anything and everything, the dual band handheld will get you going in grand style! If you are interested in propagation, and are fascinated with how radio waves bounce off the ionosphere, and get from here to there, you may wish to consider a high frequency transceiver with included 6 meters, 2 meters and 440 MHz bands. On VHF and UHF (2 meters and 440 MHz), that big mobile or base station will most likely have single sideband capabilities as well for these VHF/UHF frequencies. SSB lets you work hundreds of miles and farther, called ***"weak signal" work***. When you switch over to SSB from FM, you'll also need to switch antennas – almost all weak signal work is accomplished with an ***antenna that is horizontally polarized***. **ANSWER C.**

T3A04 What can happen if the antennas at opposite ends of a VHF or UHF line of sight radio link are not using the same polarization?

A. The modulation sidebands might become inverted.
B. Signals could be significantly weaker.
C. Signals have an echo effect on voices.
D. Nothing significant will happen.

Almost all repeater stations throughout the world use vertical polarization for transmit and receive. This means you must use the same polarization of your handheld antenna, whether it be a tiny rubber duck or a long element flexible whip. Keep YOUR handheld and antenna straight up and down, perpendicular to the Earth, for best reception. If you transmit with the handheld ***antenna horizontal*** to the Earth, your received ***signal*** could be as much as ***100 times weaker*** through that distant repeater. Now you wouldn't want that, would you? **ANSWER B.**

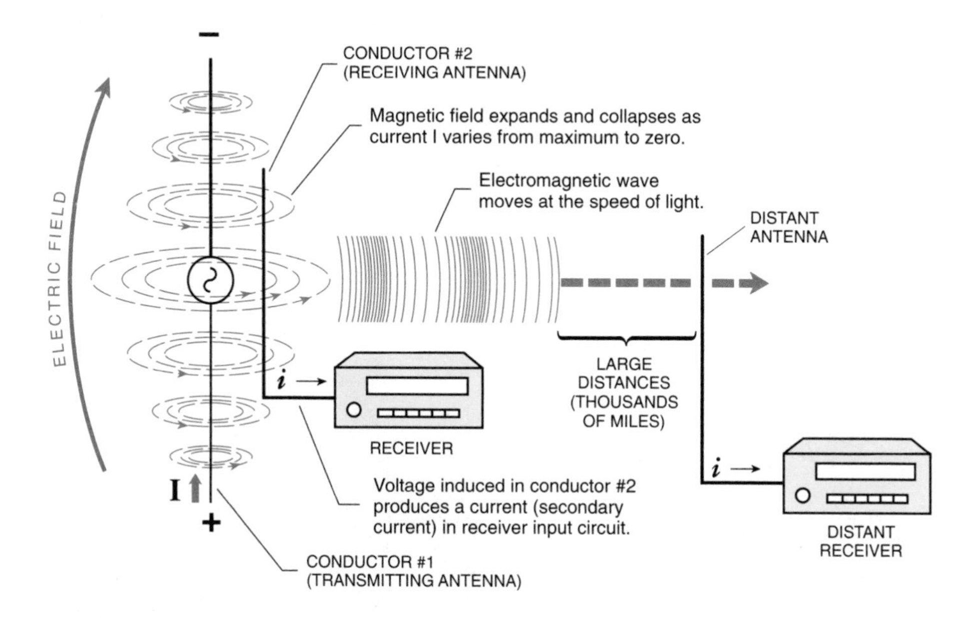

Transmitter to Receiver – Radio waves from transmitting antennas induce signals in receiving antennas as they pass by.
Source: *Listening to Shortwave,* Ken Winters, ©1993 Master Publishing, Inc., Niles, Illinois

Feed Me With Some Good Coax!

T6D11 Which of the following is a common use of coaxial cable?

A. Carry DC power from a vehicle battery to a mobile radio.
B. Carry RF signals between a radio and antenna.
C. Secure masts, tubing, and other cylindrical objects on towers.
D. Connect data signals from a TNC to a computer.

Think of coaxial cable much like a garden hose – it takes the pressure from the water company, keeps it confined down the tube, and squirts it out the other end! Same thing with coaxial cable – it ***carries your transmitted RF signal from your radio*** and delivers it ***to the antenna***. It also takes signals coming in from the antenna and delivers them to your radio – hopefully without leaks! **ANSWER B.**

T9B03 Why is coaxial cable used more often than any other feed line for amateur radio antenna systems?

A. It is easy to use and requires few special installation considerations.
B. It has less loss than any other type of feed line.
C. It can handle more power than any other type of feed line.
D. It is less expensive than any other types of feed line.

When you get started as a new Technician Class operator, most of your external antenna work will almost always include a feed line called ***coaxial cable***. Like a garden hose, don't kink it, scrunch it, chop it, or squash it in a car door. As long as you are careful to keep coax cable in its natural round shape, it ***requires few installation considerations*** and can be run right along side the vehicle frame. **ANSWER A.**

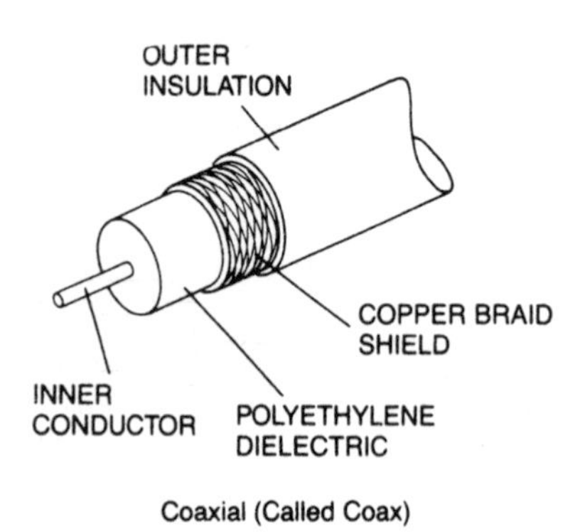

Coaxial (Called Coax)

This photo shows the different inside construction of various brands of coax cable

T9B02 What is the impedance of the most commonly used coaxial cable in typical amateur radio installations?

A. 8 ohms.
B. 50 ohms.
C. 600 ohms.
D. 12 ohms.

Most ham radio ***coax is rated at 50 ohms*** impedance. But if you accidentally slam a car door on it and distort its round shape, now the impedance will be more like 20 ohms, and this will result in a mismatch. **ANSWER B.**

T9B05 What generally happens as the frequency of a signal passing through coaxial cable is increased?

A. The apparent SWR increases.
B. The reflected power increases.
C. The characteristic impedance increases.
D. The loss increases.

The higher we go in frequency on 2 meters, 440 MHz, 900 MHz, and 1270 MHz the greater the need for larger diameter coaxial cable. ***The higher we go in frequency, the greater the loss*** of energy within coax cable. While you can get by with pencil thin round shaped coax for high frequency, I always try to use the larger cable, about the size of my thumb, for most ham radio VHF and UHF antenna installations. Tiny coax is for sissies! **ANSWER D.**

T9B07 Which of the following is true of PL-259 type coax connectors?

A. They are good for UHF frequencies.
B. They are water tight.
C. The are commonly used at HF frequencies.
D. They are a bayonet type connector.

The common ***PL-259 connector*** is what we normally use on ***high frequency radios***, and is common on your little mobile and base 2 meter/440 MHz radio, too. Never try to screw them in to an N connector receptacle! **ANSWER C.**

BNC, Type N. and PL 259 Connectors

T9B06 Which of the following connectors is most suitable for frequencies above 400 MHz?

A. A UHF (PL-259/SO-239) connector.
B. A Type N connector.
C. An RS-213 connector.
D. A DB-23 connector.

If you plan to ***operate above 400 MHz***, such as working satellites on 70 cm, or at 1270 MHz, the antennas will usually accept only a ***type N connector***. If you try to screw in a common PL-259 connector to a type N receiver, you will damage the antenna connection for life. Always look carefully to make sure you know the difference between type N connector and the more common PL-259. Type N above 400 MHz, always! **ANSWER B.**

T7C11 What is a disadvantage of "air core" coaxial cable when compared to foam or solid dielectric types?

A. It has more loss per foot.
B. It cannot be used for VHF or UHF antennas.
C. It requires special techniques to prevent water absorption.
D. It cannot be used at below freezing temperatures.

Your buddy down the street has a very special gift for you – air core coaxial cable. Tell him thanks, but no thanks. This is professional cable for commercial communications, and requires special connectors plus dry nitrogen gas feeding down the inside, and it's like working with a frozen garden hose. Stick with regular ham radio quality coaxial cable, and you will be set! That ***air core line*** is good, but ***requires special techniques to keep the water out***. **ANSWER C.**

Large coax, with hollow center conductor, low loss.

T7C09 Which of the following is the most common cause for failure of coaxial cables?

A. Moisture contamination.
B. Gamma rays.
C. The velocity factor exceeds 1.0.
D. Overloading.

Moisture getting into coax cable is the number one cause of a lousy signal on the air. Keep your coax dry. Always seal exposed coaxial cable connectors up at the antenna feedpoint. Regular coax PL259 connectors are not water-tight. **ANSWER A.**

T9B08 Why should coax connectors exposed to the weather be sealed against water intrusion?

A. To prevent an increase in feed line loss.
B. To prevent interference to telephones.
C. To keep the jacket from becoming loose.
D. All of these choices are correct.

Ok, you have that brand new antenna on the roof, and you are wearing your safety glasses and hard hat for roof safety. Good job. Before you step back to admire your work, double check for the edge of the roof, and also double check that you have completely ***sealed*** that PL-259 connector with a gooey substance called Coax Seal. This will keep the water out and ***prevent feed line loss***. **ANSWER A.**

T7C10 Why should the outer jacket of coaxial cable be resistant to ultraviolet light?

A. Ultraviolet resistant jackets prevent harmonic radiation.
B. Ultraviolet light can increase losses in the cable's jacket.
C. Ultraviolet and RF signals can mix together, causing interference.
D. Ultraviolet light can damage the jacket and allow water to enter the cable.

Always buy professional coaxial cable. Quality coax has an outer jacket that will not break down in ***ultraviolet light*** from the sun. If you try to use old CB radio coax, the jacket may already be ***breaking down, allowing moisture to enter the cable***. You'll be off the air in an instant when the next rain hits! **ANSWER D.**

T9B10 What electrical difference exists between the smaller RG-58 and larger RG-8 coaxial cables?

A. There is no significant difference between the two types.
B. RG-58 cable has less loss at a given frequency.
C. RG-8 cable has less loss at a given frequency.
D. RG-58 cable can handle higher power levels.

RG-58 coaxial cable is for sissies – it's very small and losses are high. ***RG-8U coax*** is the "big stuff" with ***less loss at any frequency***. We call it RG-8 "style" coax as there are many similar sizes of this coax with even better internal construction to minimize loss. **ANSWER C.**

Coax Cable Type, Size and Loss per 100 Feet

Coax Type	Size	Loss at HF 100 MHz	Loss at UHF 400 MHz
RG-58U	Small	4.3 dB	9.4 dB
RG-8X	Medium	3.7 dB	8.0 dB
RG-8U	Large	1.9 dB	4.1 dB
RG-213	Large	1.9 dB	4.5 dB
Hardline	Large, Rigid	0.5 dB	1.5 dB

T9B11 Which of the following types of feed line has the lowest loss at VHF and UHF?

A. 50-ohm flexible coax.
B. Multi-conductor unbalanced cable.
C. Air-insulated hard line.
D. 75-ohm flexible coax.

If you are getting real serious about operating with satellites – especially those in an elliptical orbits thousands of miles out – you may wish to locate that guy down the street who had a spool of air-insulated hardline coax. Ask him for a hunk of it and let him do the connection, too. ***Air-insulated hardline is the ultra-ultra best*** for weak signal work on VHF and UHF, but requires a big deal in getting all the connectors properly soldered. **ANSWER C.**

T7C02 Which of the following instruments can be used to determine if an antenna is resonant at the desired operating frequency?

A. A VTVM.
B. An antenna analyzer.
C. A "Q" meter.
D. A frequency counter.

Most 2 meter/440 MHz mobile and white fiberglass base antennas are pre-set to the bands of operation. Down on the worldwide bands, like your voice privileges on 10 meters, you may need to do some antenna adjustments. Handy equipment is the ***antenna analyzer***. This allows you to work on the antenna safely, on the roof, to then test its resonant frequency without needing to climb down the ladder to transmit on your big ham rig. The little SWR analyzer hooks directly to the base of the antenna and allows you to do the test easily while on the roof. **ANSWER B.**

T7C03 What, in general terms, is standing wave ratio (SWR)?

A. A measure of how well a load is matched to a transmission line.
B. The ratio of high to low impedance in a feed line.
C. The transmitter efficiency ratio.
D. An indication of the quality of your station's ground connection.

If you decide to home brew your own antenna system, this will be fun! But you'll need to obtain some additional test equipment, like a standing wave ratio (SWR) meter. This will allow you to test ***how well*** your antenna, which we call ***the load, is matched to the transmitter impedance*** as measured in your coax feed line. We would want both to be 50 ohms for a 1:1 impedance match. **ANSWER A.**

T9B01 Why is it important to have a low SWR in an antenna system that uses coaxial cable feed line?

A. To reduce television interference .
B. To allow the efficient transfer of power and reduce losses.
C. To prolong antenna life.
D. All of these choices are correct.

If your antenna match is perfect, you will have a low standing wave ratio, and there will be an ***efficient transfer of power*** up the coax to your antenna system, ***with minimum losses***. **ANSWER B.**

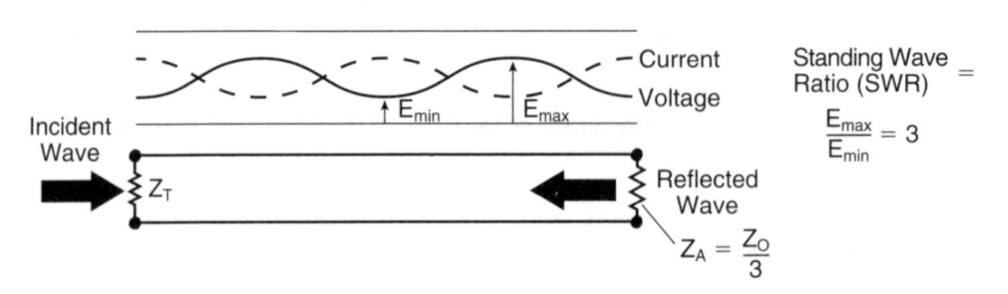

Impedance Mismatch Causes Reflected Wave

T7C04 What reading on an SWR meter indicates a perfect impedance match between the antenna and the feed line?

A. 2 to 1.
B. 1 to 3.
C. 1 to 1.
D. 10 to 1.

The ***perfect match***, as read on a SWR meter, would be ***1 to 1***. **ANSWER C.**

*SWR Reading	Antenna Condition
1:1	Perfectly Matched
1.5:1	Good Match
2:1	Fair Match
3:1	Poor Match
4:1	Something Definitely Wrong

*Constant Frequency

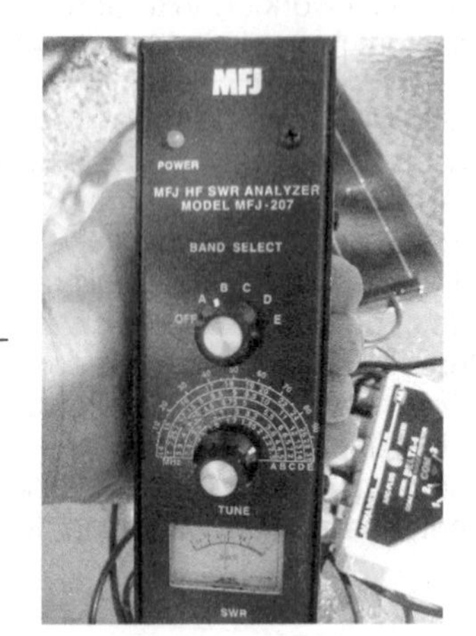

A battery operated SWR analyzer for tower antenna work.

T7C05 What is the approximate SWR value above which the protection circuits in most solid-state transmitters begin to reduce transmitter power?

A. 2 to 1.
B. 1 to 2.
C. 6 to 1.
D. 10 to 1.

When you transmit into an external antenna with your handheld, you will notice that the radio chassis will get warm. This is normal on high power after about 45 seconds of transmitting. You would be cautioned never to use high power for continuous transmitting for more than a couple of minutes or else the unit gets VERY WARM! Handheld radios do not possess the circuitry to self-protect on overheat or high SWR. On a mobile single-band or dual-band radio, they, too, will begin to get warm on transmit; and if a bad antenna is detected with an elevated SWR, you will actually see the power output indicator on the mobile begin to "fold back." This "fold back" is automatic circuitry that limits power output to protect the solid-state transmitter device. It will normally ***reduce power with a SWR higher than 2 to 1***. **ANSWER A.**

T7C06 What does an SWR reading of 4:1 mean?

A. An antenna loss of 4 dB.
B. A good impedance match.
C. An antenna gain of 4.
D. An impedance mismatch.

A ***4:1 reading*** is a sure indication that ***something is wrong*** with your antenna system. Here, the SWR surely indicates you won't get many contacts. Find out what's wrong and correct it. **ANSWER D.**

T9B09 What might cause erratic changes in SWR readings?

A. The transmitter is being modulated.
B. A loose connection in an antenna or a feed line.
C. The transmitter is being over-modulated.
D. Interference from other stations is distorting your signal.

It's a windy day today, and you notice that your local repeater jumps up and down in signal strength. In fact, you tune around and notice that many other stations are doing the same thing on your external chimney-mount, dual-band antenna. Guess what? You probably have a ***loose connection up at the antenna or*** a loose coax cable ***feed line connector***. Don't transmit – safely get up on the roof and see what is loose. **ANSWER B.**

Make sure all coax connections are tight to help minimize interference.

T7C08 What instrument other than an SWR meter could you use to determine if a feed line and antenna are properly matched?

A. Voltmeter.
B. Ohmmeter.
C. Iambic pentameter.
D. Directional wattmeter.

The technical ham may substitute an in-line, ***directional watt meter*** for an SWR meter. The most popular is manufactured by Bird, and a rotating sensor element allows the technical ham to compute power forward, and reverse power coming back due to an improperly constructed antenna or a bad feed line. **ANSWER D.**

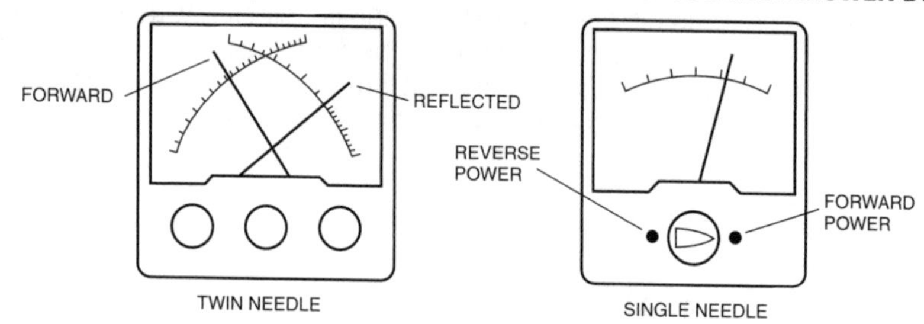

Directional Wattmeter

T7C07 What happens to power lost in a feed line?

A. It increases the SWR.
B. It comes back into your transmitter and could cause damage.
C. It is converted into heat.
D. It can cause distortion of your signal.

If moisture creeps into an exposed coax cable connector in wet weather, some of your transmitted ***energy goes up into worthless heat*** at the connection point. Keep coaxial cable connectors covered, and make sure there is no nick in the outside jacket. **ANSWER C.**

T9B04 What does an antenna tuner do?

A. It matches the antenna system impedance to the transceiver's output impedance.
B. It helps a receiver automatically tune in weak stations.
C. It allows an antenna to be used on both transmit and receive.
D. It automatically selects the proper antenna for the frequency band being used.

An ***antenna tuner*** will ***match*** your 6- and 10-meter ***transceiver to an antenna system*** that might not be perfectly tuned to the frequency on which you wish to operate. Antenna tuners are not normally found in 2-meter / 440 MHz systems. **ANSWER A.**

T7A07 If figure T5 represents a transceiver in which block 1 is the transmitter portion and block 3 is the receiver portion, what is the function of block 2?

A. A balanced modulator.
B. A transmit-receive switch.
C. A power amplifier.
D. A high-pass filter.

Here's a new Figure to gaze upon – Block 1, within your handheld is the transmit section, and Block 3 is the receiver – so guess what *Block* 2 is? That's called the ***transmit/receive switch***, and it could be a relay, or it could be a pin diode. It toggles the antenna between your transmitter and your receiver when you press or release the push-to-talk transmit button on the microphone. **ANSWER B.**

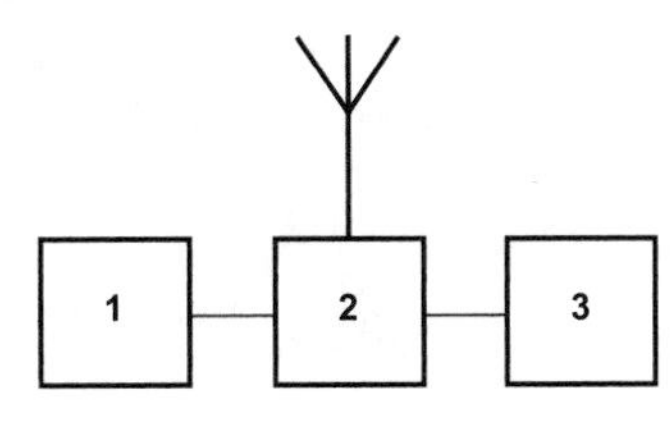

Figure T5

T7C01 What is the primary purpose of a dummy load?

A. To prevent the radiation of signals when making tests.
B. To prevent over-modulation of your transmitter.
C. To improve the radiation from your antenna.
D. To improve the signal to noise ratio of your receiver.

I like to test my brand new radios with a ***dummy load***, which ***prevents the signal from being sent out*** on the airwaves. The dummy load lets me look at my signal on a spectrum analyzer to make sure it's absolutely clean before I hook up to an outside antenna. **ANSWER A.**

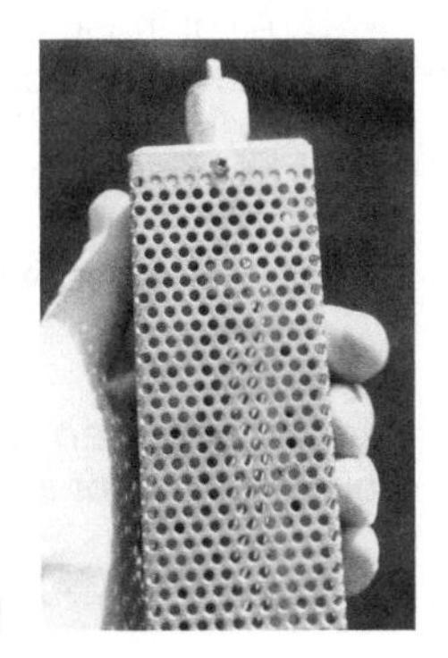

A Dummy Load

Website Resources

IF YOU'RE LOOKING FOR	THEN VISIT
Roll Your Own Antenna	www.cebik.com
Portable Satellite Beams	www.arrowantennas.com
VHF/UHF Antennas	www.AORUSA.com
	www.C3IUSA.com
	www.directivesystems.com
	www.MFJEnterprises.com
	www.M2INC.com
	www.NEW-Tronics.com
	www.DiamondAntenna.com
	www.Cushcraft.com
	www.Cubex.com
	www.Radioworks.com

CD 4 TRACK 5

Safety First!

T0A06 What is a good way to guard against electrical shock at your station?

A. Use three-wire cords and plugs for all AC powered equipment.
B. Connect all AC powered station equipment to a common safety ground.
C. Use a circuit protected by a ground-fault interrupter.
D. All of these choices are correct.

All of these are good wiring techniques that will keep you safe around some of your larger ham radio equipment as you upgrade all the way to Amateur Extra Class. **ANSWER D.**

T0A03 What is connected to the green wire in a three-wire electrical AC plug?

A. Neutral.
B. Hot.
C. Safety ground.
D. The white wire.

Remember ***green for ground***. The 3-wire electrical plug should always have the ground connector in place to provide safety. **ANSWER C.**

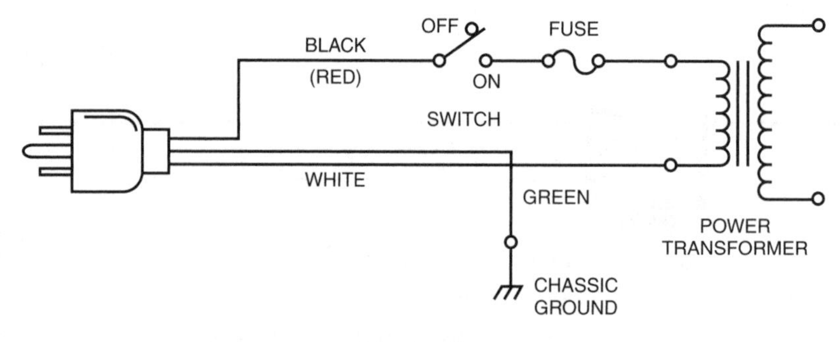

AC Line Connections

T0A13 What safety equipment should always be included in home-built equipment that is powered from 120V AC power circuits?

A. A fuse or circuit breaker in series with the AC "hot" conductor.
B. An AC voltmeter across the incoming power source.
C. An inductor in series with the AC power source.
D. A capacitor across the AC power source.

If someone gives you an old AC ham radio set, or you decide to resurrect an old ham radio "boat anchor," make sure it has ***a fuse in series with the AC hot conductor***. If the equipment is very old, the big electrolytic capacitors inside are probably bad and you want the fuse to blow first before this "boat anchor" blows up in your ham shack! **ANSWER A.**

T0A04 What is the purpose of a fuse in an electrical circuit?

A. To prevent power supply ripple from damaging a circuit.
B. To interrupt power in case of overload.
C. To limit current to prevent shocks.
D. All of these choices are correct.

Fuses provide an on-purpose weak link in case too much current passes through. The weak ***fuse*** element ***will melt*** at a specific current, and the ***circuit is now interrupted***. **ANSWER B.**

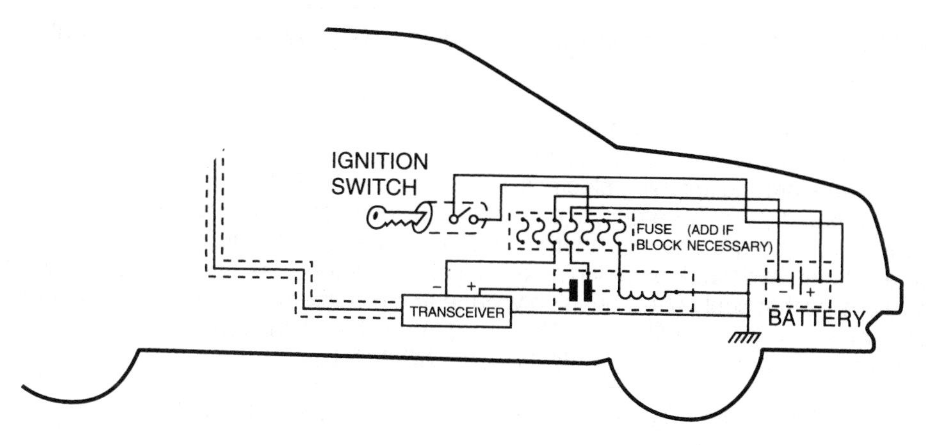

Place the fuses as close to the battery as possible

T0A05 Why is it unwise to install a 20-ampere fuse in the place of a 5-ampere fuse?

A. The larger fuse would be likely to blow because it is rated for higher current.
B. The power supply ripple would greatly increase.
C. Excessive current could cause a fire.
D. All of these choices are correct.

A fuse is installed in both the red and black power leads of your transceivers to protect the wires from overload. Pulling too much current through the wires could cause them to heat up, and for the insulation to give off toxic smoke and eventually ***burst into flames***. If you ***substitute a 20-amp fuse for a 5-amp fuse*** on a small radio that is malfunctioning and blowing the 5-amp fuse, the 20-amp fuse might carry the load, causing the wires to heat up and, POOF, you smoke your installation. **ANSWER C.**

T0A12 What kind of hazard might exist in a power supply when it is turned off and disconnected?

A. Static electricity could damage the grounding system.
B. Circulating currents inside the transformer might cause damage.
C. The fuse might blow if you remove the cover.
D. You might receive an electric shock from stored charge in large capacitors.

High-voltage ***capacitors*** will many times ***store a charge*** for several minutes after the equipment has turned off. Keep your fingers well away from any power supply until you have metered it for lethal voltages still present, even though it has been unplugged from the wall for many hours. **ANSWER D.**

T0A01 Which is a commonly accepted value for the lowest voltage that can cause a dangerous electric shock?

A. 12 volts.
C. 120 volts.
B. 30 volts.
D. 300 volts.

30 volts is dangerous! Even a couple of golf cart batteries could kill you if you aren't careful. This is why you must be especially careful not to touch any bare wires or connections when leaning across a bank of batteries because that would allow current to flow through your body accidentally. **ANSWER B.**

T0A02 How does current flowing through the body cause a health hazard?

A. By heating tissue.
B. It disrupts the electrical functions of cells.
C. It causes involuntary muscle contractions.
D. All of these choices are correct.

Never work around anything electrical in your bare feet. Put you shoes on! Never do your laundry with bare feet. If current is allowed to enter your finger and pass through your body out to ground from your bare sweaty feet, you will be toasted for sure. Never, ever work on anything electrical in bare feet! Remove those heavy gold chains that could short-out high voltage and give them to me for safe keeping! Look at ***answers A, B, and C***. Disgusting, aren't they? Indeed, current flowing through your body is a lethal health hazard! **ANSWER D.**

T0A08 What is one way to recharge a 12-volt lead-acid station battery if the commercial power is out?

A. Cool the battery in ice for several hours.
B. Add acid to the battery.
C. Connect the battery to a car's battery and run the engine.
D. All of these choices are correct.

Many ham operators power an emergency grab-and-go bag radio system from a small, rechargeable, 12-volt battery. If normal AC commercial power is out, you could still ***charge that battery by turning on your vehicle***, and then plugging into the accessory receptacle. **ANSWER C.**

T0A10 What can happen if a lead-acid storage battery is charged or discharged too quickly?

A. The battery could overheat and give off flammable gas or explode.
B. The voltage can become reversed.
C. The "memory effect" will reduce the capacity of the battery.
D. All of these choices are correct.

Overcharging any type of battery, including your little handheld battery, could cause it to ***heat up*** and ***give off dangerous gas***, or even ***explode***. Be very cautious even with your new handheld battery and only charge it from the supplied charger or from a recommended ham radio battery charger company. **ANSWER A.**

T0A09 What kind of hazard is presented by a conventional 12-volt storage battery?

A. It emits ozone which can be harmful to the atmosphere.
B. Shock hazard due to high voltage.
C. Explosive gas can collect if not properly vented.
D. All of these choices are correct.

Many ham operators who work in emergency communications have a spare 12-volt battery outside their ham shack. They can tie into it for emergency power if necessary. The reason we leave the *12-volt battery* OUTSIDE is it contains dangerous acid that could spill, has plenty of power that could cause a fire, and if improperly charged could give off an *explosive gas*. Keep those 12-volt batteries safely outside. **ANSWER C.**

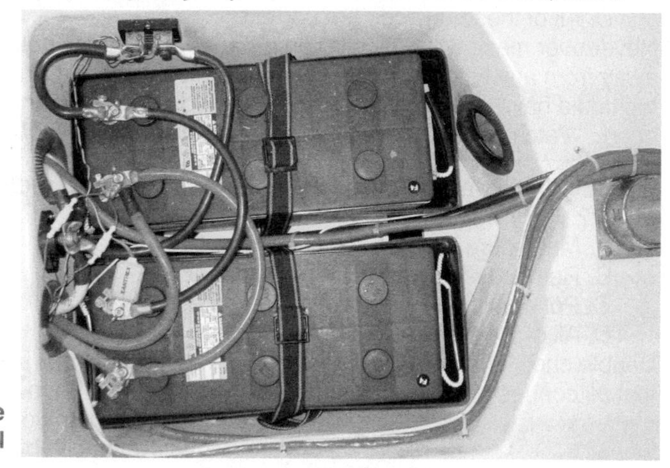

Marine storage batteries, in parallel

T0B04 Which of the following is an important safety precaution to observe when putting up an antenna tower?

A. Wear a ground strap connected to your wrist at all times.
B. Insulate the base of the tower to avoid lightning strikes.
C. Look for and stay clear of any overhead electrical wires.
D. All of these choices are correct.

Anytime you're working on an antenna system, look all over the place for any overhead electrical wires, and steer clear of these fatal attraction conductors. Normally those wires on the top of a power pole are uncovered, bare wire, and a brush with those 2 top wires will be your last days in ham radio. *Watch out for overhead electrical wires*! **ANSWER C.**

T0B06 What is the minimum safe distance from a power line to allow when installing an antenna?

A. Half the width of your property.
B. The height of the power line above ground.
C. 1/2 wavelength at the operating frequency.
D. So that if the antenna falls unexpectedly, no part of it can come closer than 10 feet to the power wires.

Make sure your antenna system cannot accidentally fall close to or on power lines. A downed antenna system should be at least *10 feet away from any nearby wires*. Your first antenna is likely to be a dual-band, white fiberglass 10' lightweight mast antenna. But you don't need to mast mount this antenna – as long as you can get it up on the roof, normally affixed to a vent pipe; you are good to go with exquisite range. You do not need to buy a bunch of extension poles to get it way up in the air – you just need to get it up and out for great results. Run good coax, the size of your thumb. No sissy stuff the size of your little finger. Attach this antenna to your two-band portable or mobile rig downstairs and enjoy some great range. **ANSWER D.**

T0B09 Why should you avoid attaching an antenna to a utility pole?

A. The antenna will not work properly because of induced voltages.
B. The utility company will charge you an extra monthly fee.
C. The antenna could contact high-voltage power wires.
D. All of these choices are correct.

Never ever mess with any wire on a phone pole, utility pole, or power pole. Stay away from any type of ***utility pole*** as likely the ***high voltage wires*** on the very top could have a little bit of leakage, and your day, and equipment, will be ruined for good. Watch out for high voltage power wires! **ANSWER C.**

T0B02 What is a good precaution to observe before climbing an antenna tower?

A. Make sure that you wear a grounded wrist strap.
B. Remove all tower grounding connections.
C. Put on a climbing harness and safety glasses.
D. All of the these choices are correct.

Double check that your ***fall prevention safety harness*** is OSHA-approved and that all connections are cinched up before climbing the tower. Always have a fellow ham help you from ground level. Never climb a tower alone! Wear ***safety goggles and a hard-hat***, and always observe climbing safety every step of the way up and down. Always double up on your safety straps once you begin working on your antenna system up on the tower. Also make sure that everyone down below wears a hard hat and safety goggles. You, too, up the tower, need a hard hat!
ANSWER C.

Always wear a hard hat and safety glasses when working on an antenna tower.

T0B03 Under what circumstances is it safe to climb a tower without a helper or observer?

A. When no electrical work is being performed.
B. When no mechanical work is being performed.
C. When the work being done is not more than 20 feet above the ground.
D. Never.

Any time you are working on the roof or on an antenna tower, make sure you have your helpers and observers adorned with plastic eye protectors (glasses), as well as hard hats. ***Never work*** on anything electrical or mechanical, or more than 2 inches off the ground, ***without a helper*** nearby. **ANSWER D.**

T0B07 Which of the following is an important safety rule to remember when using a crank-up tower?

A. This type of tower must never be painted.
B. This type of tower must never be grounded.
C. This type of tower must never be climbed unless it is in the fully retracted position.
D. All of these choices are correct.

All my towers here in Southern California are crank up. I crank them down before any big windstorm. I also ***crank them all the way down*** before I do any antenna work – never, ever climb a crank-up tower if it is partially or fully cranked up. Crank it down! Crank up towers rely on the stainless steel cable to keep them at their "ultimate high." If you climb a crank up fully extended, you could overload the old cable, and if the cable snaps, both you and the upper sections of the tower will nosedive to the ground. Never ever climb a crank up when it is still cranked up! **ANSWER C.**

If you climbed a crank-up tower and it broke and collapsed, you could lose an arm or a a leg. Crank it down!

T0B11 Which of the following establishes grounding requirements for an amateur radio tower or antenna?

A. FCC Part 97 Rules.
B. Local electrical codes.
C. FAA tower lighting regulations.
D. Underwriters Laboratories' recommended practices.

Check with ***local electrical codes*** before considering any type of amateur radio tower or antenna system. Do you antenna work when your neighbors are not around. Always wear a hard hat and safety glasses. The glasses especially if there are others working with antenna projectiles. It is always the local codes that will spell heartache or success when setting up a new antenna system in the neighborhood. If your neighborhood has no visible antennas, consider mounting your white fiberglass, two-band antenna on the back side of the roof. It will work quite nicely, even though it doesn't stick above the peak of the roof. Just get it up in the air and enjoy. Do your roof work when your neighbors are at work. You can even paint the antenna green or brown before putting it up. Go stealth, and they won't even know you are on the air with an outside antenna! **ANSWER B.**

T0B08 What is considered to be a proper grounding method for a tower?

A. A single four-foot ground rod, driven into the ground no more than 12 inches from the base.
B. A ferrite-core RF choke connected between the tower and ground.
C. Separate eight-foot long ground rods for each tower leg, bonded to the tower and each other.
D. A connection between the tower base and a cold water pipe.

Each tower leg – usually three – ***gets its own 8' ground rod***. Each rod is bonded to each other forming a triangle, then bonded to the tower for lightning protection. **ANSWER C.**

T4A08 Which type of conductor is best to use for RF grounding?

A. Round stranded wire.
B. Round copper-clad steel wire.
C. Twisted-pair cable.
D. Flat strap.

The best method of ***grounding*** your equipment is with flat copper foil ribbon. This is available from most ham radio stores. The ***flat*** ribbon offers the best surface area to bleed off static and to minimize ground currents that could cause interference. The ***strap*** usually comes in 3 inch widths, and you can fold it once or twice in order to snake it down to a healthy ground rod. **ANSWER D.**

Copper Foil Ground Strap Provides
Good Surface Area Ground

T0A11 Which of the following is good practice when installing ground wires on a tower for lightning protection?

A. Put a loop in the ground connection to prevent water damage to the ground system.
B. Make sure that all bends in the ground wires are clean, right angle bends.
C. Ensure that connections are short and direct.
D. All of these choices are correct.

When you really go gung ho with your ham radio hobby you'll want to install a ham radio tower. Watch out for high-voltage electrical wires – usually the top 2 wires are not protected with an outside insulation. Down at the base of the tower, make sure that your ***grounding wires are as short as possible and go directly to the Earth***. **ANSWER C.**

T0B10 Which of the following is true concerning grounding conductors used for lightning protection?

A. Only non-insulated wire must be used.
B. Wires must be carefully routed with precise right-angle bends.
C. Sharp bends must be avoided.
D. Common grounds must be avoided.

When you run your ***grounding conductors, avoid sharp bends***, as you don't want to confuse a lightning bolt on how to get from here to there directly. NO bends. **ANSWER C.**

T0A07 Which of these precautions should be taken when installing devices for lightning protection in a coaxial cable feed line?

A. Include a parallel bypass switch for each protector so that it can be switched out of the circuit when running high power.
B. Include a series switch in the ground line of each protector to prevent RF overload from inadvertently damaging the protector.
C. Keep the ground wires from each protector separate and connected to station ground.
D. Ground all of the protectors to a common plate which is in turn connected to an external ground.

You can buy inexpensive ***lightning protectors*** that go in series with your coaxial cable feed line. These protectors must be ***grounded to an external ground*** in order to be effective. They may save your equipment from a nearby lightning strike. If lightning should actually strike your antenna system, kiss everything goodbye for good! **ANSWER D.**

T0B01 When should members of a tower work team wear a hard hat and safety glasses?

A. At all times except when climbing the tower.
B. At all times except when belted firmly to the tower.
C. At all times when any work is being done on the tower.
D. Only when the tower exceeds 30 feet in height.

Local hams may invite you to an "antenna party." As a group, you may help another ham put the antenna up atop a tower. If you are on the ground, or up on the tower, make sure you wear a hard hat and safety glasses – ***falling*** hammers, screw drivers, and antenna ***parts-and-pieces can cause serious injury***. **ANSWER C.**

T0B05 What is the purpose of a gin pole?

A. To temporarily replace guy wires.
B. To be used in place of a safety harness.
C. To lift tower sections or antennas.
D. To provide a temporary ground.

Next time you are at an antenna raising party, you will probably see ***a gin pole***. This is a pole with a pulley that straps on to the tip top of the highest tower section, allowing you to ***lift more tower sections*** up as well as the final antenna assembly. It is a lifting device. **ANSWER C.**

T0C04 What factors affect the RF exposure of people near an amateur station antenna?

A. Frequency and power level of the RF field.
B. Distance from the antenna to a person.
C. Radiation pattern of the antenna.
D. All of these choices are correct.

The term "RF" refers to "radiofrequency" energy that comes off of your ham radio antenna system on transmit. The radiofrequency energy can be analyzed as a radiofrequency "field," and these "RF fields" must be evaluated to determine the environmental effects that this radio energy may have on our well-being. The important factors in evaluating the environmental effects of RF emissions are: frequencies and power levels of your transmitter; how high up your antenna is and how far it is away from people; the radiation pattern of your antenna; and, to some extent, ground reflections of the radiated energy. ***All of these answers*** are important to consider. **ANSWER D.**

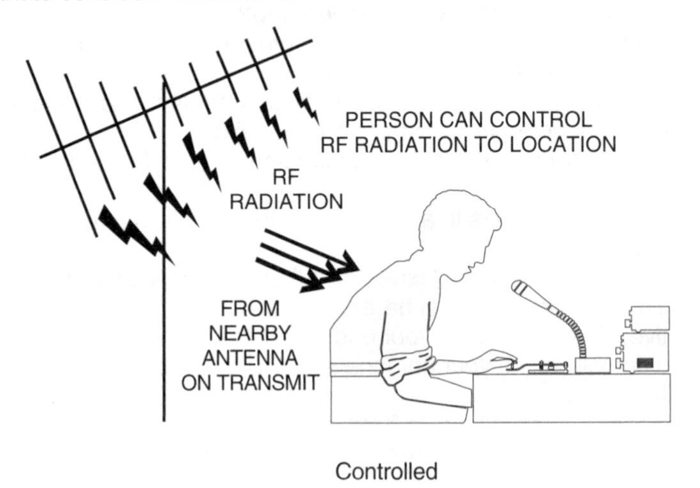

Controlled

Uncontrolled

RF Radiation Exposure Environments

T0C05 Why do exposure limits vary with frequency?

A. Lower frequency RF fields have more energy than higher frequency fields.
B. Lower frequency RF fields do not penetrate the human body.
C. Higher frequency RF fields are transient in nature.
D. The human body absorbs more RF energy at some frequencies than at others.

Remember that your operation on many different radio bands changes the frequency of your RF emissions. ***At certain wavelengths*** around 6 meters, ***the human body absorbs more RF energy***. On lower-frequency worldwide wavelengths, the body absorbs less energy. **ANSWER D.**

T0C02 Which of the following frequencies has the lowest Maximum Permissible Exposure limit?

A. 3.5 MHz.
B. 50 MHz.
C. 440 MHz.
D. 1296 MHz.

We call "maximum permissible exposure" MPE. Every radio frequency has different effects on the body, but the common Technician Class ham radio 6 meter band, at ***50 MHz, requires added distance between you and the transmitting antenna***. As long as your 6 meter antenna is more than 20 feet away from you, at modest power levels, you're going to live to a ripe old age without problems from your radio hobby. Never put any type of ham radio antenna within 5 feet of your smiling face or body. **ANSWER B.**

T0C03 What is the maximum power level that an amateur radio station may use at VHF frequencies before an RF exposure evaluation is required?

A. 1500 watts PEP transmitter output.
B. 1 watt forward power.
C. 50 watts PEP at the antenna.
D. 50 watts PEP reflected power.

At ***50 watts*** of power or more, you are required to complete a routine RF radiation evaluation. This includes VHF equipment operated by the Technician Class licensee. [97.13(C)(1)] **ANSWER C.**

Never stand in front of a microwave feedhorn antenna. On transmit, it radiates a concentrated beam of RF energy.

T0C01 What type of radiation are VHF and UHF radio signals?

A. Gamma radiation.
B. Ionizing radiation.
C. Alpha radiation.
D. Non-ionizing radiation.

The transmission of radiofrequency energy from your antenna is considered ***"nonionizing" radiation***. This is altogether different than X-ray, gamma ray, and ultra violet radiation. **ANSWER D.**

T0C06 Which of the following is an acceptable method to determine that your station complies with FCC RF exposure regulations?

A. By calculation based on FCC OET Bulletin 65.
B. By calculation based on computer modeling.
C. By measurement of field strength using calibrated equipment.
D. All of these choices are correct.

You can determine how your station complies with FCC RF exposure regulation by using The W5YI RF Safety Tables in the Appendix of this book. You can use the tables to estimated safe distances based on FCC OET Bulletin No. 65, or by your own calculations based on computer modeling. You also can actually go out there and measure with a field-strength meter the power density levels making sure to use calibrated equipment. *All of these choices* are a good way to determine whether or not you are going to expose yourself or your neighbors unnecessarily. **ANSWER D.**

T0C08 Which of the following actions might amateur operators take to prevent exposure to RF radiation in excess of FCC-supplied limits?

A. Relocate antennas.
B. Relocate the transmitter.
C. Increase the duty cycle.
D. All of these choices are correct.

In order to prevent exposure to RF radiation in excess of the FCC-specified limits, you may need to *relocate your antenna*. **ANSWER A.**

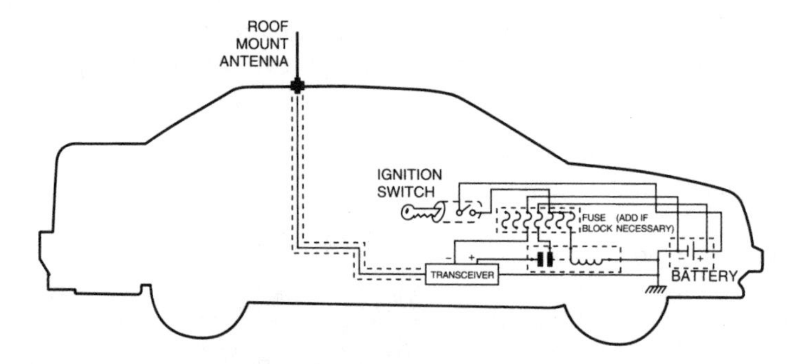

The safest place to mount the mobile antenna for minimum RF exposure is on the metal roof, as indicated.

T0C09 How can you make sure your station stays in compliance with RF safety regulations?

A. By informing the FCC of any changes made in your station.
B. By re-evaluating the station whenever an item of equipment is changed.
C. By making sure your antennas have low SWR.
D. All of these choices are correct.

As you develop your den into a full-fledged radio room, always evaluate and *re-evaluate all of the equipment* going in and all of the antennas above you, on the roof. Make sure you are not going to be transmitting and subjecting your family and neighbors to too much RF radiation. Follow guidelines we have for you, in this book, and stay safe around radio waves. **ANSWER B.**

T0C11 What is meant by "duty cycle" when referring to RF exposure?

A. The difference between lowest usable output and maximum rated output power of a transmitter.
B. The difference between PEP and average power of an SSB signal.
C. The ratio of on-air time to total operating time of a transmitted signal.
D. The amount of time the operator spends transmitting.

Think of RF exposure *"duty cycle"* to your friendly microwave oven. The oven is only "friendly" if you keep your body parts OUT of it during the time it's turned on. If you are reheating something in the microwave for *30 seconds at 50% power*, it has a lower duty cycle than cooking up some hot chocolate at *2 minutes, 100% power*. Sending data and FM has a higher duty cycle than, for instance, doing Morse code or operating SSB. The important thing is — stay the heck away from any transmitting antenna! **ANSWER C.**

T0C10 Why is duty cycle one of the factors used to determine safe RF radiation exposure levels?

A. It affects the average exposure of people to radiation.
B. It affects the peak exposure of people to radiation.
C. It takes into account the antenna feed line loss.
D. It takes into account the thermal effects of the final amplifier.

Radiofrequency emissions from your new ham radio station will vary in ***"duty cycle"*** depending on what type of emission you are using. On FM, every ***time*** you key your microphone, ***your transmitter is working*** at full tilt. On single sideband, every time you pause between words and syllables, transmitter output drops to almost zero, decreasing the overall duty cycle. When sending CW, your dots and dashes will be active energy, with spaces in between as no energy. This will also affect the duty cycle. **ANSWER A.**

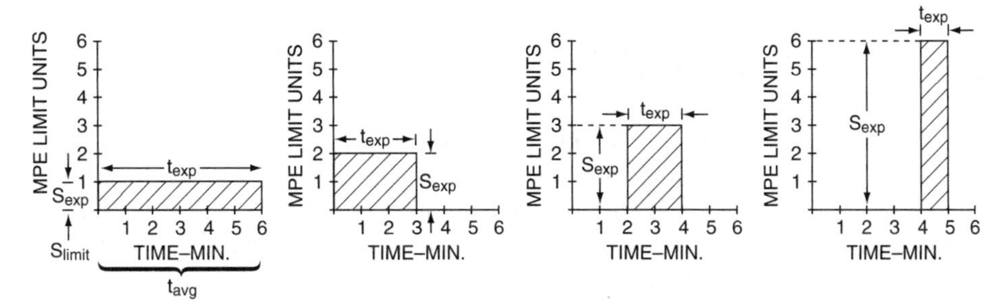

The general equation for time averaging exposure equivalence is:

$$S_{exp}\, t_{exp} = S_{limit}\, t_{avg}$$

The duty cycle is:

$$\text{DC (in \%)} = \frac{t_{exp}}{t_{avg}} \times 100$$

T0C07 What could happen if a person accidentally touched your antenna while you were transmitting?

A. Touching the antenna could cause television interference.
B. They might receive a painful RF burn.
C. They might develop radiation poisoning.
D. All of these choices are correct.

When you pass your Technician Class license exam, you're allowed to operate up to 1500 watts output of RF energy. You might run this amount of power while conducting moon-bounce operations on the 2-meter or 430-MHz band. Make sure that no one can touch the transmitting antenna with this major amount of power on it, because ***accidentally touching the antenna will cause someone to be burned*** and injured.
ANSWER B.

Be sure to place your antenna where no one can touch it!

5

Taking the Exam & Receiving Your First License

This chapter tells you when and where to test, how the examination will be given, who is qualified to give the Element 2 exam, and what happens after you complete it. There's also some good tips on finding a Volunteer Examiner team near you.

EXAMINATION ADMINISTRATION

All amateur radio service examinations are conducted by licensed amateur operators who volunteer and who are accredited by a Volunteer Examiner Coordinator (VEC). Each exam session is coordinated by a national or regional VEC. Licensed amateurs who hold a General Class or higher class license may be accredit by a VEC to administer the Element 2 Technician Class examination. Advanced Class VEs may administer Elements 2 and 3 only. Extra Class VEs may administer all examinations, including Element 4.

A team of 3 officially-accredited Volunteer Examiners (VEs) are required to form an examination session. No one-on-one or one-on-two. Three examiners must be present for the exam session to be valid.

The VEs who will conduct your exam session are your fellow hams, and as volunteers don't get a penny for their time and skills. However, they are permitted to charge you a fee for certain reimbursable expenses incurred in preparing, administering, and processing the examination. The FCC adjusts the fee annually based on inflation. The current fee is about $15.00.

HOW TO FIND AN EXAM SITE

Exam sessions are held regularly at sites throughout the country to serve their local communities. The exam site could be a public library, a fire house, someone's office, in a warehouse, and maybe even in someone's private home. Each examination team may regularly post their examination locations down at the local ham radio store. They also inform their VEC when and where they regularly hold test sessions. So the easiest way for you to find an exam session that is near you and at a convenient time is to call the VEC.

A complete list of VECs is located on page 197 in the Appendix. The W5YI VEC and the ARRL VEC are the 2 largest examination groups in the country, and they test in all 50 states. Their 3-member, accredited examination teams are just about

Want to find a test site fast?
Visit the W5YI-VEC website at **www.w5yi.org**, or call 800-669-9594.

everywhere. So when you call the W5YI-VEC in Texas, or the ARRL-VEC in Connecticut, be assured they probably have an examination team only a few miles from where you are reading this book right now!

Any of the VECs listed will provide you with the phone number of a local Volunteer Examiner team leader who can tell you the schedule of upcoming exam sessions near you. Go ahead — give them a call now and let them know you are studying my book. Select a session you wish to attend, and make a reservation so the VEs know to expect you. Don't be a "no-show" and don't be a "surprise-show." Make a reservation. And don't hesitate to tell them how much we all appreciate their efforts in supporting ham radio testing.

Ask them ahead of time what you will need to bring to the examination session. And ask them how much the current fee is for your exam session.

Remember, your volunteer examiners don't get paid for their time, so anything that you can do to help out during the exam session will be appreciated. Maybe stick around after the exam session to help the VEs put away the tables and chairs. Someday *you* may be a volunteer examiner, and you will appreciate the help!

WHAT TO BRING TO THE EXAM

Here's what you'll need to bring with you for your Technician Class, Element 2 written examination:

1. Examination fee of approximately $15.00 in cash.
2. Personal identification with a photo.
3. Any Certificates of Successful Completion of Examinations (CSCEs) issued within the last 365 days prior to this test session date. Bring the originals, plus two copies of everything.
4. Some sharp pencils and fine-tip pens. Bring a backup!
5. Calculators may be used, so bring your calculator.
6. Any other item that the VE team asks you to bring. Remember, these volunteer examiners receive no pay for their work.

EXAM CONTENT

Years ago, the FCC staff administered amateur radio exams, and the questions and answers were *secret.* The FCC developed their secret questions, their secret multiple-choice answers, and they had all sorts of secret subjects that you never knew about until after you took the exam for the first time.

In the 80's, things improved when President Ronald Reagan signed legislation providing for volunteer amateur operator examinations, allowing the Federal Communications Commission to transfer testing responsibilities over to the amateur community. This included making up the examination questions and answers, which would then be available in the public domain. It is the role of Volunteer Exam Coordinators — specifically the Question Pool Committee — to review the questions in each of the three pools for the various amateur operator licenses. This procedure has been in effect for years for FAA airplane pilot testing. The amateur radio exam process has been privatized, and has worked out very well.

So, there won't be any surprises on you upcoming Element 2 written examination. Again, every one of the 35 questions on your exam will be taken from the 394 question pool in this book. The wording of the questions, answer, and distracters will be exactly as they appear here. The only change that the VECs are allowed to make is in the order of the A B C D answers.

TAKING THE EXAMINATION

Get a good night's sleep before exam day. Continue to study theory Q&A up to the moment you go into the room. Make a list of questions you have the most difficulty answering. Memorize the answers, and review them as you go to the examination. If someone is going with you to the examination, have them ask you the questions on the way, so that you can practice answering them. Listen to my theory CDs in your car as you drive to the examination session. Speedread ***keywords*** over and over again before the exam!

Check and Double-Check

When the examiners hand out the examination material, put your name, date, and test number on the answer sheet. *Make no marks on the multiple-choice question sheet.* Only write on the answer sheets.

Read over the examination questions carefully. Take your time in looking for the correct answer. Some answers start out looking correct, but end up wrong. Don't speed read the test.

When you are finished with the examination, go back over every question and double-check your answers. Try a game where you read what you have selected as the correct answer, and see if it agrees with the question.

When you are finished with the exam, turn in all of your paperwork. Tell the examiners how much you appreciate their efforts to help promote ham radio participation. If you are the last one in the room, volunteer to help them take down the testing location. The VE team will appreciate your offer.

And now, wait patiently outside for the examiners to announce that you have passed your examination. Chances are they will greet you with a smile and your Certificate of Successful Completion of Examination. Make sure to immediately sign this certificate when it is handed to you.

COMPLETING NCVEC FORM 605

When you arrive at the examination site, one of the first things you will do is complete the NCVEC Form 605. This form is retained by the Volunteer Exam Coordinator who transfers your printed information to an electronic file and sends it to the FCC for your new license, or upgrade. Your application may be delayed or kicked-back to you if the VEC can't read your writing. Make absolutely sure you print as legibly as you can, and carefully follow the instructions on the form.

Name

This isn't a tough one — your last name, first name, middle initial, and suffix such as junior or senior. You must stay absolutely consistent with your name on any future Form 605s for upgrades or changes of address. If you start out as "Jack" and end up "John," the computer will throw out your next application. If you don't list a middle initial the first time, but do the second time, the computer will again hiccup. If you decide to use a nickname, this is okay — but down the line when you visit a foreign country, they may ask you for your personal identification that needs to illustrate this same nickname. It is best to stick with the name that is on most of your personal pictured IDs, such as your Driver's License.

NCVEC QUICK-FORM 605 APPLICATION FOR AMATEUR OPERATOR/PRIMARY STATION LICENSE

SECTION 1 - TO BE COMPLETED BY APPLICANT

PRINT LAST NAME	SUFFIX	FIRST NAME	INITIAL	STATION CALL SIGN (IF ANY)
MARCONI		JOE	G	

MAILING ADDRESS (Number and Street or P.O. Box)	SOCIAL SECURITY NUMBER / TIN (OR LICENSEE ID)
7101 RECTIFIER ROAD	090-909-090

CITY	STATE CODE	ZIP CODE (5 or 9 Numbers)	E-MAIL ADDRESS (OPTIONAL)
INDUCTOR	IL	60777	SPARKS@MSN.COM

DAYTIME TELEPHONE NUMBER (Include Area Code) OPTIONAL	FAX NUMBER (Include Area Code) OPTIONAL	ENTITY NAME (IF CLUB, MILITARY RECREATION, RACES)

Type of Applicant: [X] Individual [ ] Amateur Club [ ] Military Recreation [ ] RACES (Renewal Only)

TRUSTEE OR CUSTODIAN CALL SIGN

SIGNATURE OF RESPONSIBLE CLUB OFFICIAL

I HEREBY APPLY FOR (Make an X in the appropriate box(es))

- [X] **EXAMINATION** for a **new** license grant
- [ ] **EXAMINATION** for **upgrade** of my license class
- [ ] **CHANGE** my **name** on my license to my new name

  Former Name: ______________ (Last name) (Suffix) (First name) (MI)
- [ ] **CHANGE** my mailing address to **above** address
- [ ] **CHANGE** my station **call sign** systematically

  Applicant's Initials: ______________
- [ ] **RENEWAL** of my license grant.

Do you have another license application on file with the FCC which has not been acted upon?	PURPOSE OF OTHER APPLICATION	PENDING FILE NUMBER (FOR VEC USE ONLY)

I certify that:

- I waive any claim to the use of any particular frequency regardless of prior use by license or otherwise;
- All statements and attachments are true, complete and correct to the best of my knowledge and belief and are made in good faith;
- I am not a representative of a foreign government;
- I am not subject to a denial of Federal benefits pursuant to Section 5301of the Anti-Drug Abuse Act of 1988, 21 U.S.C. § 862;
- The construction of my station will NOT be an action which is likely to have a significant environmental effect (See 47 CFR Sections 1.301-1.319 and Section 97.13(a));
- I have read and WILL COMPLY with Section 97.13(c) of the Commission's Rules regarding RADIOFREQUENCY (RF) RADIATION SAFETY and the amateur service section of OST/OET Bulletin Number 65.

Signature of applicant (Do not print, type, or stamp. Must match applicant's name above.)

X Joe G Marconi   Date Signed: 4-15-00

NCVEC Form 605

Social Security Number

Put in your Social Security Number in the designated box. If you are a citizen of another country, put down the country name in this box. If you are a U.S. citizen and prefer not to disclose your Social Security Number on this application, you will want to check with the VEC ahead of time and follow their detailed instructions on how to secure an FCC "CORES FRN ID" number to use in lieu of your SSN. Take our word for it — just give them your SSN, and you will avoid a lot of grief.

Address

Where do you want your paper license mailed? If you move around frequently, you will need to be contacting the FCC regularly for a change of address. Use a mailing address that you plan to keep as permanent as possible.

E-mail Address

This is optional, but it's a good idea because the amateur radio service is now under the FCC's Universal Licensing System. Once you get your new call sign, you will be able to work with the Federal Communications Commission directly via computer, including change of address, change of name, and license renewals without having to do any paperwork.

Phone Numbers

There are two boxes for phone numbers — one for a daytime contact, and the other for your FAX number. These are optional, but it's a good idea to put these numbers down just in case the VEC or VE team need to re-contact you for some reason.

Signature

When you sign your name, make sure to include all of the letters that you printed as your name at the top of the form. Don't just put down a squiggle or an initial. You need to sign your name all the way out, including all of the letters that were in your printed name.

Final Check

Finally, double-check that your handwriting is legible. If a single letter in your name can't be read clearly and is misinterpreted, subsequent electronic filings may get returned as no action. Make sure your NCVEC Form 605 is as clear as a bell to your Volunteer Examination team, who will then forward it to their VEC. Stay away from red ink, too.

Your Examiners' Portion

The Volunteer Examination Team will carefully review your NCVEC Form 605 to ensure that they can read your handwriting and that everything looks okay. Then VEs will sign and date your form and send it on to the VEC for processing. The VEC will then electronically file your results with the FCC.

CONGRATULATIONS! YOU PASSED!

After you pass the examination, congratulations and a big welcome to Technician Class privileges are in order! The world of microwave and VHF/UHF operating awaits you. And welcome to the additional HF privileges and the world of long-range, high-frequency operation that you earned.

You can begin operating as soon as you obtain your official FCC call sign. This means you do not have to wait for the hard copy of your license to arrive at the address listed on your NCVEC 605.

Electronic Filing

After you pass your Technician Class examination, your VE team will submit your amateur license application results electronically to their Volunteer Examiner Coordinator (VEC). The VEC will then electronically file for your license grant, and within days of passing your test your call sign is granted and you are permitted to go on the air immediately!

The system of electronic filing has been working so well that most new applicants who pass their exam over the weekend are able to go on the air that following Wednesday or Thursday once they have seen their new call letters appear on the FCC database. Recent rule changes now allow you to instantly go on the air with those call letters, as an FCC grant, even though you don't actually have the official FCC paper license in your possession. It takes about two weeks to receive that official FCC license, which you may wish to frame, or cut off the top portion to keep in your wallet.

Ask your Volunteer Examination team when they will electronically file your test results. They can then give you an approximate date to check the FCC database to find out your new call sign.

YOUR FIRST CALL SIGN

To find your new call sign on the internet, go to: **http://wireless.fcc.gov/**, which will take you to the FCC Wireless Telecommunications Bureau home page. On the right hand side of the screen under Licensing click on "License Search" then search by name entering your last name first, a comma, and then your first name. You also can look for your new call sign on www.qrz.com.

Amateur Call Signs

As an aid to enforcement of the radio rules, transmitting stations throughout the world are required to identify themselves at regular intervals when they are in operation. That's the primary purpose of your call sign. A call sign is a very important matter to a ham — sometimes more personal than his or her name!

By international agreement, the prefix letters of a station's call sign indicates the country in which that station is authorized to operate. The national prefixes allocated to the United States are AA through AL, KA through KZ, NA through NZ, and WA through WZ. In addition, U.S. amateur stations with call signs that start with AA-AL, K, N or W are followed by a number indicating their location in a specific U.S. geographic area. *Table 5-1* details these geographical areas. On the DX airwaves, hams can readily identify the national origin of the ham signal they hear by its call sign prefix. The suffix letters indicate a specific amateur station.

The Amateur Operator/Primary Station call sign is issued by the FCC's licensing facility in Gettysburg, Pennsylvania, on a systematic basis after it receives your application information from the VEC. There are four call sign groupings, A, B, C, and D. Tables showing the call sign groups and formats for operator/station licenses within and outside the contiguous U.S. are shown in the Appendix.

- Group A call signs are issued to Extra Class licensees, and have a 1-by2, 2-by-1, or 2-by-2 format that begins with the letter A. W9MU is an example.
- Group B call signs are issued to grandfathered Advanced Class licensees, and have a 2-by-2 format that begins with K, N, or W. WA6PT is an example.
- Group C call signs are issued to General, grandfathered Technician-Plus, and Technician class licensees, and have a 1-by-3 format. N7LAT is an example. (However, because of the large number of Technician licenses that have been issued since the inception of the "no-code" license in 1991, there are no Group C call signs available. Thus, in accordance with FCC rules, new General and Technician class operators are issued a Group D call sign.)
- Group D call signs are issued to new Technician and General class operators, and to grandfathered Novice class operators. They have a 2-by-3 format. KB9SMG is an example.

Table 5-1. Call Sign Area for U.S. Geographical Areas

Call Sign Area No.	Geographical Area
1	Maine, New Hampshire, Vermont, Massachusetts, Rhode Island, Connecticut
2	New York, New Jersey, Guam and the U.S. Virgin Islands
3	Pennsylvania, Delaware, Maryland, District of Columbia
4	Virginia, North and South Carolina, Georgia, Florida, Alabama, Tennessee, Kentucky, Midway Island, Puerto Rico[1]
5	Mississippi, Louisiana, Arkansas, Oklahoma, Texas, New Mexico
6	California, Hawaii[2]
7	Oregon, Washington, Idaho, Montana, Wyoming, Arizona, Nevada, Utah, Alaska[3]
8	Michigan, Ohio, West Virginia, American Samoa
9	Wisconsin, Illinois, Indiana
0	Colorado, Nebraska, North and South Dakota, Kansas, Minnesota, Iowa, Missouri, Northern Mariana Island

[1] Puerto Rico also issued Area #3.
[2] Hawaii also issued Area #7.
[3] Alaska also issued Areas #1 through #0.

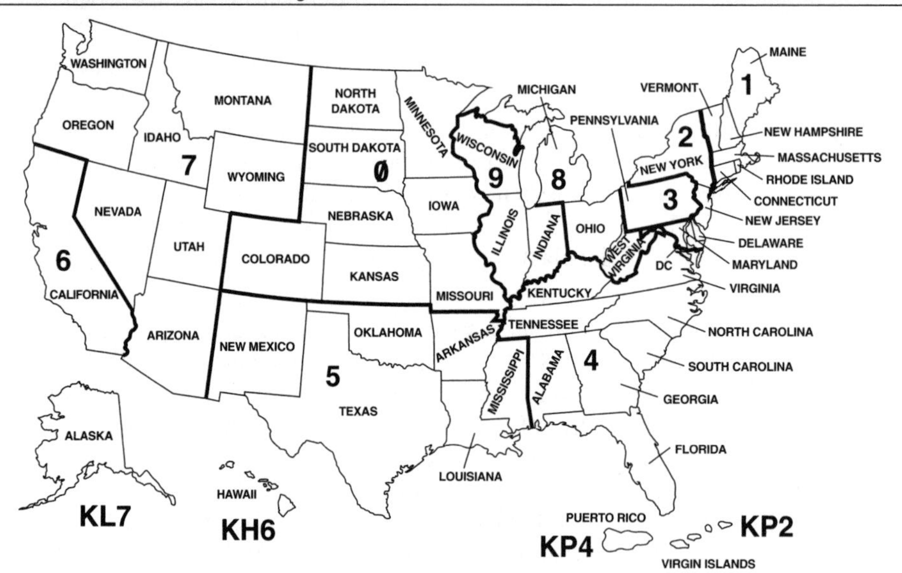

Figure 5-1. U.S. Call Sign Areas

Once assigned, a call sign is never changed unless the licensee specifically requests the change — even if they move out of the continental United States. FCC licensees residing in foreign countries must show a U.S. mailing address on their applications. You may change your call sign to a new group when you upgrade your amateur license.

Of course, if you get a real neat call sign as your first issued call sign, you can elect to hold onto it all the way to the top. My very first call sign was WB6NOA, and I've retained this same call sign all the way up to my current Extra Class license.

So what does all this mean? You can no longer easily tell what grade of license someone has by how many letters and single number in their call sign. But you're just a few keystrokes away from finding out on your computer by visiting **www.qrz.com** on the worldwide web. This website contains a data base of call signs that you can scan. You can also look up call signs by the name of the licensee, or by the person's call sign Go to: **http://wireless.fcc.gov**. On the right side of the screen under "Licensing" click on "License Search" and then search either by name or by call sign.

The W5YI VEC also offers several call sign services to amateurs. They can handle address changes and license renewals for you. In addition, they can obtain a one-by-one call sign for your special event, or do the research and provide evidence that you were licensed prior to March 21, 1987 and therefore qualify for a General Class ticket without further examination. W5YI VEC also can file for your club station call sign. Check their website at **www.w5yi.org** for more information on these and other services it offers, or call 800-669-9594 during regular business hours.

Vanity Call Signs

Once you receive your first "no-choice," computer-assigned call letters, you may be eligible to replace them with a vanity call sign of your choosing. This call sign could be made up of your initials, or represent your love of animals (K9DOG) or could be call letters that your late mom or dad had when they got started in ham radio years ago.

General and Technician class amateur operators may request a vanity call sign from Group D or Group C. However, since Group C call signs beginning with the letter N are completely used up, only Group D call signs are available. These call signs have two letters, a number, and three letters (such as KA5GMO).

You also may request a call sign that was previously assigned to you that may have expired years ago, as well as a call sign of a close relative or former holder who is now deceased. This call sign can be from any group. There is an additional fee for a vanity call sign.

There are two ways to file for a vanity call sign — electronically or in writing.

You can file FCC Form 605 and pay the filing fee using your credit card electronically. Instructions for filing electronically are on the FCC website at **http://wireless.fcc.gov/services/amateur/callsigns/vanity**.

To file in writing, you must complete FCC Form 605 along with Form 159 and include your check or money order and mail it to: FCC, PO Box 358130, Pittsburgh, PA 15251-5130.

To make it easier for you to select a vanity call sign, you may wish to contact the W5YI Group at 800-669-9594 and ask for their vanity call sign application. The W5YI Group can help you to file electronically to get the exact call sign of your choice.

As for me, I'm staying with my original-issue WB6NOA call sign. If I changed it, I would be breaking a 50-year tradition!

THANK YOUR VE TEAM!

Your Volunteer Examination team is made up of men and women who are volunteering their time to provide you with an examination opportunity. Your VE team typically spends three additional hours for every one hour at the actual examination site. Since most exam sessions last for three hours, your three examiners may be spending as much as nine additional hours in electronic processing of your paperwork, handling the paperwork, sending the results to their Volunteer Examiner Coordinators, paying for the examination room, and all of the other chores that go along with conducting a volunteer exam session.

Work closely with your VE team and follow their instructions specifically. They are the absolute boss at the exam session, and you must follow their instructions to the letter.

Keep in mind that the volunteers are not paid. The fee that you are charged to take the exams is used to pay expenses. Your Volunteer Examiners don't keep any part of this fee to pay for their time in volunteering their services. So let them know you are grateful that they have given up their weekend or evening to provide you with an examination opportunity.

When you achieve the General Class level, or higher, it's time for you to become a Volunteer Examiner. Ask the Volunteer Examination team how you can sign up when you make General Class, and higher.

Finally, please tell your Volunteer Examination Team how much Gordon West Radio School appreciates their testing efforts. Show them these comments in this book. Tell them that Gordon West sincerely appreciates all of the hard work they are putting in to help the amateur service grow.

YOUR NEXT STEP

First, I want you to GET ON THE AIR as a new Technician Class operator. That's where you'll really begin your education as a ham radio operator. After you've spent some time operating on repeaters, and trying out skywaves on 6- and 10-meters, it will be time to consider upgrading to General Class, and then on to top license, Amateur Extra Class. Remember, no more code test, so all you'll need to upgrade are my *General Class* and *Extra Class books*.

SUMMARY

Welcome to the new, improved and simplified Amateur Radio service. There is no longer a Morse code test for any class of license. Restructured and simplified Element 2, Technician. Straight-forward Element 3, General Class. And for you high-techies, Element 4, the Amateur Extra Class.

Learn the code. Even though it's no longer requited, never in the history of ham radio has CW given you so many operating privileges on the worldwide bands as the new General Class license.

Become an *active* amateur, and help establish new radio clubs and volunteer examination programs near where you live. Introduce ham radio to kids, and let's keep our service growing!

FREE PASSING CERTIFICATE

When you pass your exams, I want to know about it! I have a very nice certificate available to you, suitable for framing plus free operating materials from ham equipment manufacturers. All I need is a large, self-addressed envelope with 12 first-class stamps on the inside to cover postage and handling and I'll send one your way. Write me at:

Gordon West Radio School
2414 College Drive
Costa Mesa, California 92626

Listen for me on the airwaves as WB6NOA. Say "hi" at many of the hamfests that I attend throughout the country every year. And if you ever would just like to speak with me, call me Monday through Friday, 10 am to 4 pm (California time), 714-549-5000.

Welcome to ham radio! It's been **FUN** teaching you the Technician Class.

Gordon West 73

Gordon West, WB6NOA

6

Learning Morse Code

On April 15, 2000, the FCC dropped the 20- and 13-word-per-minute Morse code requirements for worldwide frequency privileges for General and Extra Class operators down to 5 words per minute. On February 23, 2007, the FCC ***totally eliminated*** the Morse code test as a prerequisite for high frequency operation.

The elimination of the Morse code test for operation on worldwide frequencies conforms to international radio regulations. In 2003, the International Telecommunications Union (ITU) World Radiocommunication Conference voted to allow individual nations to determine whether or not to retain a Morse code test as a requirement to operate on frequencies below 30 MHz.

When the FCC eliminated the Morse code test for Technician Class operators in 1991 for VHF/UHF operating, the ruling was adopted with little opposition. However, the announcement that the FCC was considering total elimination of the Morse code test drew thousands of written comments to the FCC. Many comments supported code test elimination, while a minority urged the FCC to retain a code test because of the strong tradition of CW as a ham radio operating mode.

The Federal Communications Commission concluded that "…this change (eliminating the code test) eliminates an unnecessary burden that may discourage current amateur radio operators from advancing their skills and participating more fully in the benefits of amateur radio." The FCC Commissioners recognized the simple fact that learning Morse code was keeping many very technical, talented hams from obtaining their General and Extra Class licenses. Morse code is much like musical rhythms. Some people are tone deaf, and some people couldn't carry a rhythm in a hand basket.

So for years, the Morse code test was an insurmountable hurdle to many talented Technician Class hams who wanted to upgrade. If I could take these Techs and put them into one of my regular Morse code classes, we usually could get the majority of them through the CW test with outside home study, on the air practice (on 2 meters), and classroom study followed by the code test. But throughout the country, Morse code classes were few and far between and it is tough to learn new music and a new language without classroom instruction.

Technician Class operators could not practice on the worldwide airwaves to learn the code because these bands were reserved for only those operators who had already passed the code test. Running Morse code practice on a local 2 meter repeater was one option, but nothing beats the excitement of practicing code on the worldwide bands and hooking up with another station thousands of miles away.

As of February 23, 2007, we can now take new operators and introduce them to Morse code on the exciting worldwide bands!

Morse code is the ham radio operator's most basic language of short and long sounds, dits and dahs, or dots and dashes. Sailors have pounded SOS when trapped beneath a sailboat hull. In submarines, the tapping of Morse code gets the message through when there is no other way to communicate. Prisoners of war have tapped out Morse code messages, or BLINKED the code when being publicly displayed on television.

Ham operators use the code to get through when noise would otherwise cover up data or voice signals. Years ago, before road rage, fellow hams driving might greet each other by sending on their car horns H-I, a friendly salute to another ham.

So I encourage you to learn code. It is best mastered by sound along with memorizing the Morse code patterns seen on the upcoming pages. The pages show the number of dots and dashes to learn for a specific character, and learning the sound (rhythm) of Morse code is always the best way to practice. Let's see what these short sounds and long sounds are all about.

LOOKING AT MORSE CODE

The International Morse code, originally developed as the American Morse code by Samuel Morse, is truly international — all countries use it, and most commercial worldwide services employ operators who can recognize it. It is made up of short and long duration sounds. Long sounds, called "dahs," are three times longer than short sounds, called "dits." *Figure 6-1* shows the time intervals for Morse code sounds and spaces. *Figure 6-3,* on the next page, indicates the sounds for all the CW characters and symbols.

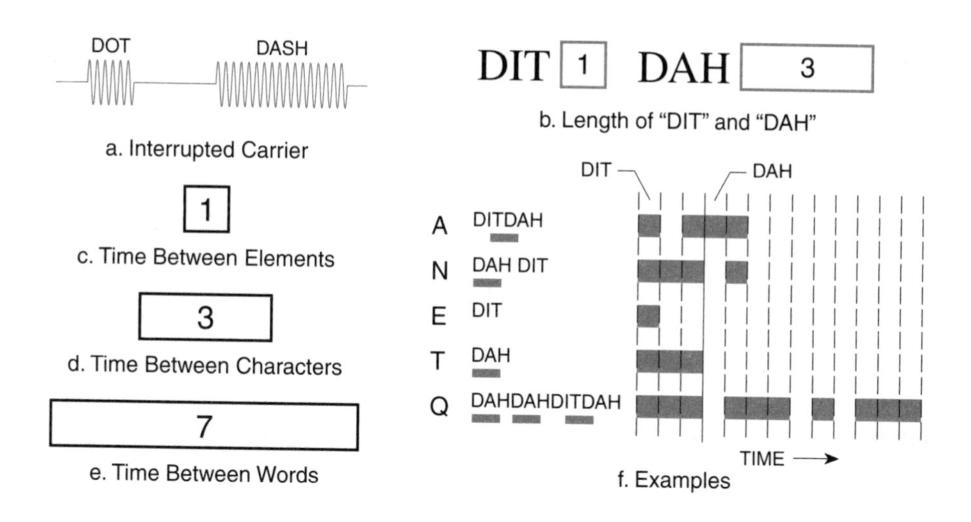

Figure 6-1. Time Intervals for Morse Code

a. Alphabet

LETTER	Composed of:	Sounds like:	LETTER	Composed of:	Sounds like:
A	• —	didah	N	— •	dahdit
B	— • • •	dahdididit	O	— — —	dahdahdah
C	— • — •	dahdidahdit	P	• — — •	didahdahdit
D	— • •	dahdidit	Q	— — • —	dahdahdidah
E	•	dit	R	• — •	didahdit
F	• • — •	dididahdit	S	• • •	dididit
G	— — •	dahdahdit	T	—	dah
H	• • • •	didididit	U	• • —	dididah
I	• •	didit	V	• • • —	didididah
J	• — — —	didahdahdah	W	• — —	ditdahdah
K	— • —	dahdidah	X	— • • —	dahdididah
L	• — • •	didahdidit	Y	— • — —	dahdidahdah
M	— —	dahdah	Z	— — • •	dahdahdidit

b. Special Signals and Punctuation

CHARACTER	Meaning:	Composed of:	Sounds like:
$\overline{AR}$	(end of message)	• — • — •	didahdidahdit
K	invitation to transmit (go ahead)	— • —	dahdidah
$\overline{SK}$	End of work	• • • — • —	dididIdahdidah
$\overline{SOS}$	International distress call	• • • — — — • • •	didididahdahdahdididit
V	Test letter (V)	• • • —	didididah
R	Received, OK	• — •	didahdit
$\overline{BT}$	Break or Pause	— • • • —	dahdididIdah
$\overline{DN}$	Slant Bar	— • • — •	dahdididahdit
$\overline{KN}$	Back to You Only	— • — — •	dahdidahdahdit
Period		• — • — • —	didahdidahdidah
Comma		— — • • — —	dahdahdididahdah
Question mark		• • — — • •	dididahdahdidit
@	For Web Address	• — — • — •	didahdahdidahdi

c. Numerals

NUMBER	Composed of:	Sounds like:
1	• — — — —	didahdahdahdah
2	• • — — —	dididahdahdah
3	• • • — —	didididahdah
4	• • • • —	dididididah
5	• • • • •	dididididit
6	— • • • •	dahdidididit
7	— — • • •	dahdahdididit
8	— — — • •	dahdahdahdidit
9	— — — — •	dahdahdahdahdit
Ø	— — — — —	dahdahdahdahdah

Figure 6-3. Morse Code and Its Sound

CODE KEY

Morse code is usually sent by using a code key. A typical one is shown in *Figure 6-2a*. Normally it is mounted on a thin piece of wood or plexiglass. Make sure that what you mount it on is thin; if the key is raised too high, it will be uncomfortable to the wrist. The correct sending position for the hand is shown in *Figure 6-2b*.

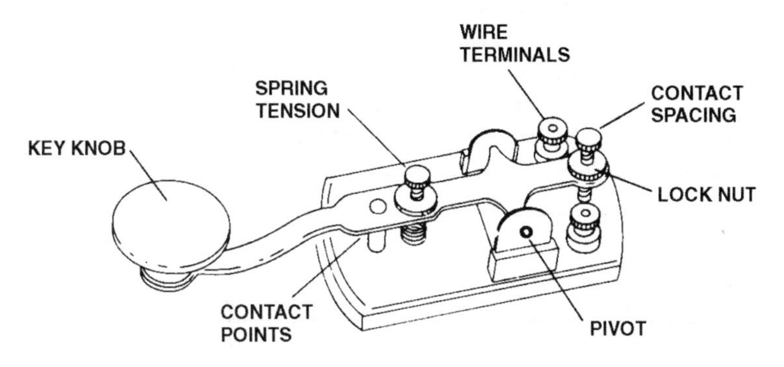

Figure 6-2a. Code Key

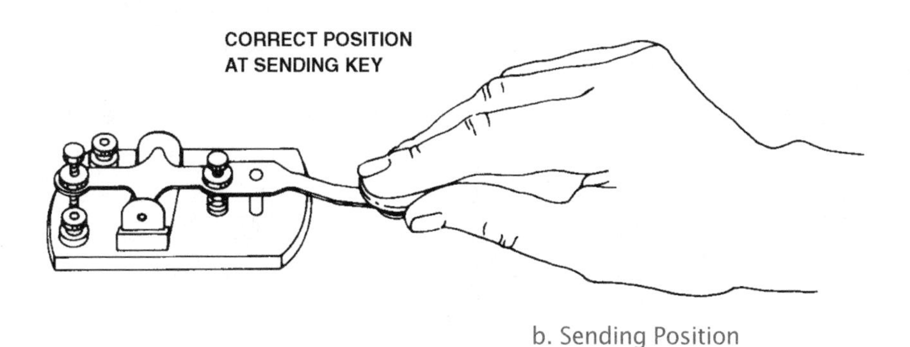

b. Sending Position

Figure 6-2b. Code Key for Sending Code

LEARNING MORSE CODE

The reason you are learning the Morse code is to be able to operate all modes on the worldwide bands—including CW. Here are five suggestions (four serious ones) on how to learn the code:

1. Memorize the code from the code charts in this book.
2. Use my fun audio course available at all ham radio stores, and from the W5YI Group.
3. Go out and spend $1,000 and buy a worldwide radio, and listen to the code live and on the air. You don't need to spend that much, but you can listen to Morse code practice on the air, as shown in *Table 6-1.*
4. Use a code key and oscillator to practice sending the code. Believe it or not, someday you're actually going to do code over the live airwaves, using this same code key hooked up to your new megabuck transceiver.
5. Play with code programs on your computer, and *have fun!*

Table 6-1. Radio Frequencies and Times for Code Reception

Pacific	Mountain	Central	Eastern	Mon.	Tue.	Wed.	Thu.	Fri.
6 a.m.	7 a.m.	8 a.m.	9 a.m.		Fast Code	Slow Code	Fast Code	Slow Code
7 a.m. – 1 p.m.	8 a.m. – 2 p.m.	9 a.m. – 3 p.m.	10 a.m. – 4 p.m.	VISITING OPERATOR TIME				
1 p.m.	2 p.m.	3 p.m.	4 p.m.	Fast Code	Slow Code	Fast Code	Slow Code	Fast Code
2 p.m.	3 p.m.	4 p.m.	5 p.m.	Code Bulletin				
3 p.m.	4 p.m.	5 p.m.	6 p.m.	Digital Bulletin				
4 p.m.	5 p.m.	6 p.m.	7 p.m.	Slow Code	Fast Code	Slow Code	Fast Code	Slow Code
5 p.m.	6 p.m.	7 p.m.	8 p.m.	Code Bulletin				
6 p.m.	7 p.m.	8 p.m.	9 p.m.	Digital Bulletin				
6:45 p.m.	7:45 p.m.	8:45 p.m.	9:45 p.m.	Voice Bulletin				
7 p.m.	8 p.m.	9 p.m.	10 p.m.	Fast Code	Slow Code	Fast Code	Slow Code	Fast Code
8 p.m.	9 p.m.	10 p.m.	11 p.m.	Code Bulletin				

CW is broadcast on the following MHz frequencies: 1.8025, 3.5815, 7.0475, 14.0475, 18.0975, 21.0675, 28.0675, and 147.555. W1AW schedule courtesy of *QST* magazine.

CODE COURSES ON CDs AND CASSETTE TAPES

Five words per minute is so slow, and so easy, that many ham radio applicants learn it completely in a single week! You can do it, too, by using the code CDs and tapes mentioned above.

Code courses personally recorded by me make code learning *fun.* They will train you to send and receive the International Morse code in just a few short weeks. They are narrated and parallel the instructions in this book. The CDs have code characters generated at a 15-wpm character rate, spaced out to a 5-wpm word rate. This is known as Farnsworth spacing.

Getting Started

The hardest part of learning the code is taking the first CD out of the case, putting it in your player, and pushing the play button! Try it, and you will be over your biggest hurdle. After that, the CDs will talk you through the code in no time at all.

The CDs make code learning *fun.* You'll hear how humor has been added to the learning process to keep your interest high. Since ham radio is a hobby, there's no reason we can't poke ourselves in the ribs and have a little fun learning the code as part of this hobby experience. Okay, you're still not convinced — you probably have already made up your mind that trying to learn the code will be the hardest part of being a ham. It will not. Give yourself a fair chance. Don't get discouraged. Have patience and remember these important reminders when practicing to learn the Morse code:

- Learn the code by sound. Don't stare at the tiny dots and dashes that we have here in the book — the dit and dah sounds on the CDs and on the air and with your practice keyer will ultimately create an instant letter at your fingertips and into the pencil.
- *Neve*r scribble down dots or dashes if you forget a letter. Just put a small dash on your paper for a missed letter. You can go back and figure out what the word is by the letters you did copy!
- Practice only with fast code characters; 15-wpm character speed, spaced down to 5-wpm speed, is ideal.
- Practice the code by writing it down whenever possible. This further trains your brain and hand to work together in a subconscious response to the sounds you hear. (Remember Pavlov and his dog "Spot"?)
- Practice only for 15 minutes at a time. The CDs will tell you when to start and when to stop. Your brain and hand will lose that sharp edge once you go beyond 16 minutes of continuous code copy. You will learn much faster with five 15-minute practices per day than a one-hour marathon at night.
- Stay on course with the cassette instructions. Learn the letters, numbers, punctuation marks, and operating signals in the order they are presented. My code teaching system parallels that of the American Radio Relay League, Boy Scouts of America, the Armed Forces, and has worked for thousands in actual classroom instruction.

It was no accident that Samuel Morse gave the single dit for the letter "E" which occurs most often in the English language. He determined the most used letters in the alphabet by counting letters in a printer's type case. He reasoned a printer would have more of the most commonly-used letters. It worked! With just the first lesson, you will be creating simple words and simple sentences with no previous background.

Table 6-2 shows the sequence of letters, punctuation marks, operating signals, and numbers covered in six lessons on the CDs recorded by me.

Table 6-2. Sequence of Lessons on Cassettes

▪ Lesson 1	E T M A N I S O SK Period
▪ Lesson 2	R U D C 5 Ø AR Question Mark
▪ Lesson 3	K P B G W F H BT Comma
▪ Lesson 4	Q L Y J X V Z DN 1 2 3 4 6 7 8 9
▪ Lesson 5	Random code with narrated answers
▪ Lesson 6	A typical 5-wpm code test

CODE KEY AND OSCILLATOR – HAM RECEIVER

All worldwide ham transceivers have provisions for a code key to be plugged in for both CW practice off the air as well as CW operating on the air. If you already own a worldwide set, chances are all you will need is a code key for some additional code-sending practice.

Read over your worldwide radio instruction manual where it talks about hooking up the code key. For code practice, read the notes about operating with a "side tone" but not actually going on the air. This "side tone" capability of most worldwide radios will eliminate your need for a separate code oscillator.

Code Key and Oscillator — Separate Unit

Many students may wish to simply buy a complete code key and oscillator set. They are available from local electronic outlets or through advertisements in the ham magazines.

Look again at the code key in *Figure 6-2a.* Note the terminals for the wires. Connect wires to these terminals and tighten the terminals so the wires won't come loose. The two wires will go either to a code oscillator set or to a plug that connects into your ham transceiver. Hook up the wires to the plug as described in your ham transceiver instruction book or the code oscillator set instruction book.

Mount the key firmly, as previously described, then adjust the gap between the contact points. With most new telegraph keys, you will need a pair of pliers to loosen the contact adjustment knob. It's located on the very end of your keyer. First loosen the lock nut, then screw down the adjustment until you get a gap no wider than the thickness of a business card. You want as little space as possible between the points. The contact points are located close to the sending plastic knob.

Now turn on your set or oscillator and listen. If your hear a constant tone, check that the right-hand movable shorting bar is not closed. If it is, swing it open. Adjust the spring tension adjustment screw so that you get a good "feel" each time you push down on the key knob. Adjust it tight enough to keep the contacts from closing while your fingers are resting on the key knob.

Pick up the key by the knob! This is the exact position your fingers should grasp the knob—one or two on top, and one or two on the side of it. Poking at the knob with one finger is unacceptable. Letting your fingers fly off the knob between dots and dashes (dits and dahs) also is not correct. As you are sending, you should be able to instantly pick up the whole key assembly to verify proper finger position.

Your arm and wrist should barely move as you send CW. All the action is in your hand — and it should be almost effortless. Give it a try, and look at *Figure6-2b* again to double-check your hand position.

Letting someone else use the key to send CW to you will also help you learn the code.

Morse Code Computer Software

The newest way to learn Morse code is through computer-aided instruction. There are many good PC programs on the market that not only teach you the characters, but build speed and allow you to take actual telegraphy examinations, which the computer constructs. Personal computer programs also can be used to make audio tapes on your tape recorder so you can listen to them on the cassette player in your car.

A big advantage of computer-aided Morse code learning is that you can easily customize the program to fit your own needs! You can select the sending speed, Farnsworth character-spacing speed, duration of transmission, number of characters in a random group, tone frequency — and more!

Some have built-in "weighting." That means the software will determine your weaknesses and automatically adjust future sending to give you more study on your problem characters! All Morse code software programs transmit the tone by keying the PC's internal speaker. Some generate a clearer audio tone through the use of external oscillators or internal computer sound cards.

Get those 6 code CDs and start listening to my voice, and see how easy it is to master the dots and dashes. Continuously push yourself to the letters with more dots and dashes in them, and work those tapes regularly and keep your copy in your spiral-bound notebook.

I hope to hear your CW call on the worldwide bands soon!

APPENDIX

U.S. VOLUNTEER EXAMINER COORDINATORS IN THE AMATEUR SERVICE

Anchorage Amateur Radio Club
PO Box 670616
Chugiak, AK 99567-0616
907/338-0662
e-mail: jwiley@alaska.net

ARRL/VEC
225 Main Street
Newington, CT 06111-1494
860/594-0300
860/594-0339 (fax)
e-mail: vec@arrl.org
Internet: www.arrl.org

Central America VEC, Inc.
Larry Frost KR4GU
2751 Christian LN NE
Huntsville, AL 35811-1864
256/288-0392
256/653-5007
e-mail: cavec@bellsouth.net

Golden Empire Amateur Radio Society
P.O. Box 508
Chico, CA 95927-0508
530/345-3515
e-mail: wa6zrt@sbcglobal.net

Greater Los Angeles Amateur Radio Group
9737 Noble Avenue
North Hills, CA 91343-2403
818/892-2068
818/892-9855 (fax)
e-mail: gla.arg@gte.net

Jefferson Amateur Radio Club
Keith Barnes W5KB
PO Box 73665
Metairie, LA 70033-3665
504/831-1613
e-mail: w5kb@w5gad.org
Internet: www.w5gad.org

Laurel Amateur Radio Club, Inc.
4708 Montgomery PL
Beltsville, MD 20705-2921
301/937-0394 (6-9 PM)
e-mail: aa3of@arrl.net
Internet: www.larcmdorg.doore.net/vec

The Milwaukee Radio Amateurs Club, Inc.
P.O. Box 070695
Milwaukee, WI 53207-0695
262/797-6722
e-mail: tom@supremecom.biz

MO-KAN/VEC
228 Tennessee Road
Richmond, KS 66080-9174
785/867-2011
e-mail: wo0e@lcwb.coop

SANDARC-VEC
P.O. Box 2446
La Mesa, CA 91943-2446
619/697-1475
e-mail: n6nyx@arrl.net

Sunnyvale VEC Amateur Radio Club, Inc.
P.O. Box 60307
Sunnyvale, CA 94088-0307
408/255-9000 (exam info 24 hours)
e-mail: vec@amateur-radio.org
Internet: www.amateur-radio.org

W4VEC
Rae Everhart K4SWN
P.O. Box 482
China Grove, NC 28023-0482
e-mail: raef@lexcominc.net
Internet: www.w4vec.com

Western Carolina WCARS
7 Skylyn Ct.
Asheville, NC 28806-3922
e-mail: bstewart@windstream.net
Internet: www.wcarsvec.org

W5YI-VEC
P.O. Box 565101
Dallas, TX 75356-5101
817/860-3800
800-669-9495
e-mail: w5yi-vec@w5yi.org
Internet: www.w5yi.org

THE W5YI RF SAFETY TABLES

(Developed by Fred Maia, W5YI, working in cooperation with the ARRL.)

There are two ways to determine whether your station's radio frequency signal radiation is within the MPE (Maximum Permissible Exposure) guidelines established by the FCC for ***"controlled"*** and ***"uncontrolled"*** environments. One way is direct ***"measurement"*** of the RF fields. The second way is through ***"prediction"*** using various antenna modeling, equations and calculation methods described in the FCC's *OET Bulletin 65* and *Supplement B.*

In general, most amateurs will not have access to the appropriate calibrated equipment to make precise field strength/power density measurements. The field-strength meters in common use by amateur operators are inexpensive, hand-held field strength meters that do not provide the accuracy necessary for reliable measurements, especially when different frequencies may be encountered at a given measurement location. It is more practical for amateurs to determine their PEP output power at the antenna and then look up the required distances to the controlled/uncontrolled environments using the following tables, which were developed using the prediction equations supplied by the FCC.

The FCC has determined that radio operators and their families are in the "controlled" environment and your neighbors and passers-by are in the "uncontrolled" environment. The estimated minimum compliance distances are in meters from the transmitting antenna to either the occupational/controlled exposure environment ("Con") or the general population/uncontrolled exposure environment ("Unc") using typical antenna gains for the amateur service and assuming 100% duty cycle and maximum surface reflection. Therefore, these charts represent the worst case scenario. They do not take into consideration compliance distance reductions that would be caused by:

(1) Feed line losses, which reduce power output at the antenna especially at the VHF and higher frequency levels.

(2) Duty cycle caused by the emission type. The emission type factor accounts for the fact that, for some modulated emission types that have a non-constant envelope, the PEP can be considerably larger than the average power. Multiply the distances by 0.4 if you are using CW Morse telegraphy, and by 0.2 for two-way SSB (single sideband) voice. There is no reduction for FM.

(3) Duty cycle caused by on/off time or "time-averaging." The RF safety guidelines permit RF exposures to be averaged over certain periods of time with the average not to exceed the limit for continuous exposure. The averaging time for occupational/controlled exposures is 6 minutes, while the averaging time for general population/uncontrolled exposures is 30 minutes. For example, if the relevant time interval for time-averaging is 6 minutes, an amateur could be exposed to two times the applicable power density limit for three minutes as long as he or she were not exposed at all for the preceding or following three minutes.

A routine evaluation is not required for vehicular mobile or hand-held transceiver stations. Amateur Radio operators should be aware, however, of the potential for exposure to RF electromagnetic fields from these stations, and take measures (such as reducing transmitting power to the minimum necessary, positioning the radiating antenna as far from humans as practical, and limiting continuous transmitting time) to protect themselves and the occupants of their vehicles.

Amateur Radio operators should also be aware that the FCC radio-frequency safety regulations address exposure to people — and not the strength of the signal. Amateurs may exceed the Maximum Permissible Exposure (MPE) limits as long as no one is exposed to the radiation.

How to read the chart: If you are radiating 500 watts from your 10 meter dipole (about a 3 dB gain), there must be at least 4.5 meters (about 15 feet) between you (and your family) and the antenna — and a distance of 10 meters (about 33 feet) between the antenna and your neighbors.

Medium and High Frequency Amateur Bands

All distances are in meters

Freq. (MF/HF) (MHz/Band)	Antenna Gain (dBi)	Peak Envelope Power (watts) 100 watts		500 watts		1000 watts		1500 watts	
		Con.	Unc.	Con.	Unc.	Con.	Unc.	Con.	Unc.
2.0 (160m)	0	0.1	0.2	0.3	0.5	0.5	0.7	0.6	0.8
2.0 (160m)	3	0.2	0.3	0.5	0.7	0.6	1.06	0.8	1.2
4.0 (75/80m)	0	0.2	0.4	0.4	1.0	0.6	1.3	0.7	1.6
4.0 (75/80m)	3	0.3	0.6	0.6	1.3	0.9	1.9	1.0	2.3
7.3 (40m)	0	0.3	0.8	0.8	1.7	1.1	2.5	1.3	3.0
7.3 (40m)	3	0.5	1.1	1.1	2.5	1.6	3.5	1.9	4.2
7.3 (40m)	6	0.7	1.5	1.5	3.5	2.2	4.9	2.7	6.0
10.15 (30m)	0	0.5	1.1	1.1	2.4	1.5	3.4	1.9	4.2
10.15 (30m)	3	0.7	1.5	1.5	3.4	2.2	4.8	2.6	5.9
10.15 (30m)	6	1.0	2.2	2.2	4.8	3.0	6.8	3.7	8.3
14.35 (20m)	0	0.7	1.5	1.5	3.4	2.2	4.8	2.6	5.9
14.35 (20m)	3	1.0	2.2	2.2	4.8	3.0	6.8	3.7	8.4
14.35 (20m)	6	1.4	3.0	3.0	6.8	4.3	9.6	5.3	11.8
14.35 (20m)	9	1.9	4.3	4.3	9.6	6.1	13.6	7.5	16.7
18.168 (17m)	0	0.9	1.9	1.9	4.3	2.7	6.1	3.3	7.5
18.168 (17m)	3	1.2	2.7	2.7	6.1	3.9	8.6	4.7	10.6
18.168 (17m)	6	1.7	3.9	3.9	8.6	5.5	12.2	6.7	14.9
18.168 (17m)	9	2.4	5.4	5.4	12.2	7.7	17.2	9.4	21.1
21.145 (15m)	0	1.0	2.3	2.3	5.1	3.2	7.2	4.0	8.8
21.145 (15m)	3	1.4	3.2	3.2	7.2	4.6	10.2	5.6	12.5
21.145 (15m)	6	2.0	4.6	4.6	10.2	6.4	14.4	7.9	17.6
21.145 (15m)	9	2.9	6.4	6.4	14.4	9.1	20.3	11.1	24.9
24.99 (12m)	0	1.2	2.7	2.7	5.9	3.8	8.4	4.6	10.3
24.99 (12m)	3	1.7	3.8	3.8	8.4	5.3	11.9	6.5	14.5
24.99 (12m)	6	2.4	5.3	5.3	11.9	7.5	16.8	9.2	20.5
24.99 (12m)	9	3.4	7.5	7.5	16.8	10.6	23.7	13.0	29.0
29.7 (10m)	0	1.4	3.2	3.2	7.1	4.5	10.0	5.5	12.2
29.7 (10m)	3	2.0	4.5	4.5	10.0	6.3	14.1	7.7	17.3
29.7 (10m)	6	2.8	6.3	6.3	14.1	8.9	19.9	10.9	24.4
29.7 (10m)	9	4.0	8.9	8.9	19.9	12.6	28.2	15.4	34.5

VHF/UHF Amateur Bands

All distances are in meters

Freq. (MF/HF) (MHz/Band)	Antenna Gain (dBi)	Peak Envelope Power (watts) 50 watts		100 watts		500 watts		1000 watts	
		Con.	Unc.	Con.	Unc.	Con.	Unc.	Con.	Unc.
50 (6m)	0	1.0	2.3	1.4	3.2	3.2	7.1	4.5	10.1
50 (6m)	3	1.4	3.2	2.0	4.5	4.5	10.1	6.4	14.3
50 (6m)	6	2.0	4.5	2.8	6.4	6.4	14.2	9.0	20.1
50 (6m)	9	2.8	6.4	4.0	9.0	9.0	20.1	12.7	28.4
50 (6m)	12	4.0	9.0	5.7	12.7	12.7	28.4	18.0	40.2
50 (6m)	15	5.7	12.7	8.0	18.0	18.0	40.2	25.4	56.8
144 (2m)	0	1.0	2.3	1.4	3.2	3.2	7.1	4.5	10.1
144 (2m)	3	1.4	3.2	2.0	4.5	4.5	10.1	6.4	14.3
144 (2m)	6	2.0	4.5	2.8	6.4	6.4	14.2	9.0	20.1
144 (2m)	9	2.8	6.4	4.0	9.0	9.0	20.1	12.7	28.4
144 (2m)	12	4.0	9.0	5.7	12.7	12.7	28.4	18.0	40.2
144 (2m)	15	5.7	12.7	8.0	18.0	18.0	40.2	25.4	56.8
144 (2m)	20	10.1	22.6	14.3	32.0	32.0	71.4	45.1	101.0
222 (1.25m)	0	1.0	2.3	1.4	3.2	3.2	7.1	4.5	10.1
222 (1.25m)	3	1.4	3.2	2.0	4.5	4.5	10.1	6.4	14.3
222 (1.25m)	6	2.0	4.5	2.8	6.4	6.4	14.2	9.0	20.1
222 (1.25m)	9	2.8	6.4	4.0	9.0	9.0	20.1	12.7	28.4
222 (1.25m)	12	4.0	9.0	5.7	12.7	12.7	28.4	18.0	40.2
222 (1.25m)	15	5.7	12.7	8.0	18.0	18.0	40.2	25.4	56.8
450 (70cm)	0	0.8	1.8	1.2	2.6	2.6	5.8	3.7	8.2
450 (70cm)	3	1.2	2.6	1.6	3.7	3.7	8.2	5.2	11.6
450 (70cm)	6	1.6	3.7	2.3	5.2	5.2	11.6	7.4	16.4
450 (70cm)	9	2.3	5.2	3.3	7.3	7.3	16.4	10.4	23.2
450 (70cm)	12	3.3	7.3	4.6	10.4	10.4	23.2	14.7	32.8
902 (33cm)	0	0.6	1.3	0.8	1.8	1.8	4.1	2.6	5.8
902 (33cm)	3	0.8	1.8	1.2	2.6	2.6	5.8	3.7	8.2
902 (33cm)	6	1.2	2.6	1.6	3.7	3.7	8.2	5.2	11.6
902 (33cm)	9	1.6	3.7	2.3	5.2	5.2	11.6	7.3	16.4
902 (33cm)	12	2.3	5.2	3.3	7.3	7.3	16.4	10.4	23.2
1240 (23cm)	0	0.5	1.1	0.7	1.6	1.6	3.5	2.2	5.0
1240 (23cm)	3	0.7	1.6	1.0	2.2	2.2	5.0	3.1	7.0
1240 (23cm)	6	1.0	2.2	1.4	3.1	3.1	7.0	4.4	9.9
1240 (23cm)	9	1.4	3.1	2.0	4.4	4.4	9.9	6.3	14.0
1240 (23cm)	12	2.0	4.4	2.8	6.2	6.2	14.0	8.8	19.8

All distances are in meters. To convert from meters to feet multiply meters by 3.28. Distance indicated is shortest line-of-sight distance to point where MPE limit for appropriate exposure tier is predicted to occur.

AUTHORIZED FREQUENCY BANDS – AMATEUR SERVICE (for U.S. Amateur Stations operating from ITU-Region 2–North and South America)

Current License Class[1]		Technician	General		Extra Class
Grandfathered[2]	Novice			Advanced	
METERS					
160			1800-2000 kHz/All	1800-2000 kHz/All	1800-2000 kHz/All
80 / 75	3525-3600 kHz/CW	3525-3600 kHz/CW	3525-3600 kHz/CW 3800-4000 kHz/Ph	3525-3600 kHz/CW 3700-4000 kHz/Ph	3500-4000 kHz/CW 3600-4000 kHz/Ph
40	7025-7125 kHz/CW	7025-7125 kHz/CW	7025-7125 kHz/CW 7175-7300 kHz/Ph	7025-7125 kHz/CW 7125-7300 kHz/Ph	7000-7300 kHz/CW 7125-7300 kHz/Ph
30			10.1-10.15 MHz/CW	10.1-10.15 MHz/CW	10.1-10.15 MHz/CW
20			14.025-14.15 MHz/CW 14.225-14.35 MHz/Ph	14.025-14.15 MHz/CW 14.175-14.35 MHz/Ph	14.0-14.35 MHz/CW 14.15-14.35 MHz/Ph
17			18.068-18.11 MHz/CW 18.11-18.168 MHz/Ph	18.068-18.11 MHz/CW 18.11-18.168 MHz/Ph	18.068-18.11 MHz/CW 18.11-18.168 MHz/Ph
15	21.025-21.2 MHz/CW	21.025-21.2 MHz/CW	21.025-21.2 MHz/CW 21.275-21.45 MHz/Ph	21.025-21.2 MHz/CW 21.225-21.45 MHz/Ph	21.0-21.45 MHz/CW 21.2-21.45 MHz/Ph
12			24.89-24.99 MHz/CW 24.93-24.99 MHz/Ph	24.89-24.99 MHz/CW 24.93-24.99 MHz/Ph	24.89-24.99 MHz/CW 24.93-24.99 MHz/Ph
10	28.0-28.5 MHz/CW 28.3-28.5 MHz/Ph	28.0-28.5 MHz/CW 28.3-28.5 MHz/Ph	28.0-28.3 MHz/CW 28.3-29.7 MHz/Ph	28.0-28.3 MHz/CW 28.3-29.7 MHz/Ph	28.0-29.7 MHz/CW 28.3-29.7 MHz/Ph
6		50-54 MHz/CW 50.1-54 MHz/Ph	50-54 MHz/CW 50.1-54 MHz/Ph	50-54 MHz/CW 50.1-54 MHz/Ph	50-54 MHz/CW 50.1-54 MHz/Ph
2		144-148 MHz/CW 144.1-148 MHz/All	144-148 MHz/CW 144.1-148 MHz/Ph	144-148 MHz/CW 144.1-148 MHz/All	144-148 MHz/CW 144.1-148 MHz/All
1.25	222-225 MHz/All	222-225 MHz/All [3]	222-225 MHz/All	222-225 MHz/All	222-225 MHz/All
0.70		420-450 MHz/All	420-450 MHz/All	420-450 MHz/All	420-450 MHz/All
0.33		902-928 MHz/All	902-928 MHz/All	902-928 MHz/All	902-928 MHz/All
0.23	1270-1295 MHz/All	1240-1300 MHz/All	1240-1300 MHz/All	1240-1300 MHz/All	1240-1300 MHz/All

[1] Effective 4-15-00 [2] Prior to 4-15-00 [3] Effective 2/1/94 219-220 MHz is authorized for point-to-point fixed digital message forwarding systems.

Note: Morse code (CW, A1A) may be used on any frequency allocated to the amateur service. Telephony emission (abbreviated Ph above) authorized on certain bands as indicated. Higher class licensees may use slow-scan television and facsimile emissions on the Phone bands; radio teletype/digital on the CW bands. All amateur modes and emissions are authorized above 144.1 MHz. In actual practice, the modes/emissions used are somewhat more complicated than shown above due to the existence of various band plans and "gentlemen's agreements" concerning where certain operations should take place.

COMMON CW ABBREVIATIONS

AA	All after
AB	All before
ABT	About
ADR	Address
AGN	Again
ANT	Antenna
BCI	Broadcast interference
BK	Break; break me; break in
BN	All between; been
B4	Before
C	Yes
CFM	Confirm; I confirm
CK	Check
CL	I am closing my station; call
CLD-CLG	Called; calling
CUD	Could
CUL	See you later
CUM	Come
CW	Continuous Wave
DLD-DLVD	Delivered
DX	Distance
FB	Fine business; excellent
GA	Go ahead (or resume sending)
GB	Good-by
GBA	Give better address
GE	Good evening
GG	Going
GM	Good morning
GN	Good night
GND	Ground
GUD	Good
HI	The telegraphic laugh; high
HR	Here; hear
HV	Have
HW	How
LID	A poor operator
MILS	Milliamperes
MSG	Message; prefix to radiogram
N	No
ND	Nothing doing
NIL	Nothing; I have nothing for you
NR	Number
NW	Now; I resume transmission
OB	Old boy
OM	Old man
OP-OPR	Operator
OT	Old timer; old top
PBL	Preable
PSE-PLS	Please
PWR	Power
PX	Press
R	Received as transmitted; are
RCD	Received
REF	Refer to; referring to; reference
RPT	Repeat; I repeat
SED	Said
SEZ	Says
SIG	Signature; signal
SKED	Schedule
SRI	Sorry
SVC	Service; prefix to service message
TFC	Traffic
TMW	Tomorrow
TNX	Thanks
TU	Thank you
TVI	Television interference
TXT	Text
UR-URS	Your; you're; yours
VFO-	Variable-frequency oscillator
VY	Very
WA	Word after
WB	Word before
WD-WDS	Word; words
WKD-WKG	Worked; working
WL	Well; will
WUD	Would
WX	Weather
XMTR	Transmitter
XTAL	Crystal
XYL	Wife
YL	Young lady
73	Best regards
88	Love and kisses

SCHEMATIC SYMBOLS

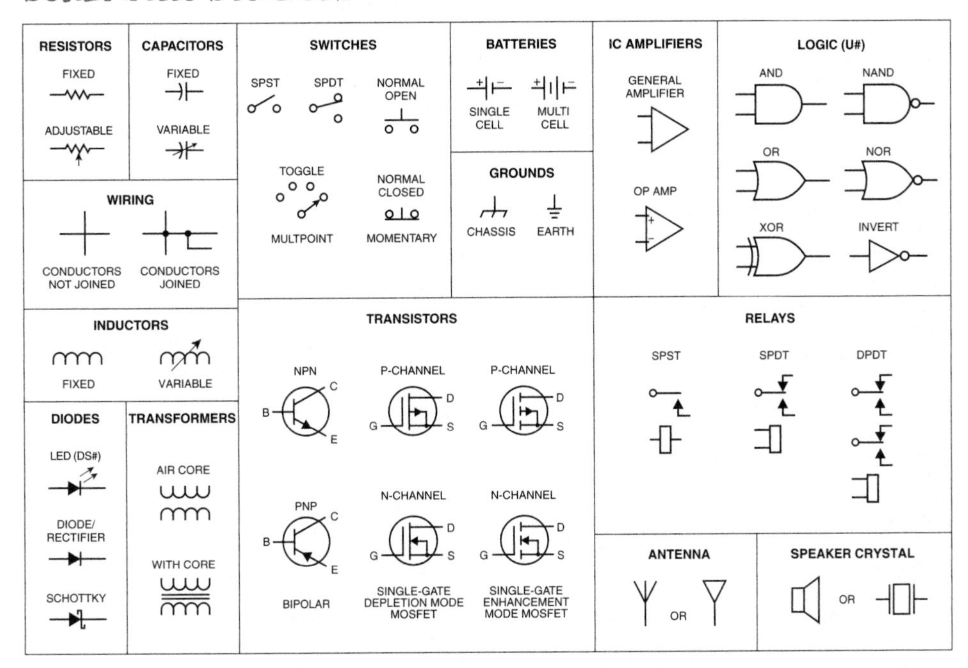

SCIENTIFIC NOTATION

Prefix	Symbol	Multiplication Factor
exa	E	10^{18} = 1,000,000,000,000,000,000
peta	P	10^{15} = 1,000,000,000,000,000
tera	T	10^{12} = 1,000,000,000,000
giga	G	10^{9} = 1,000,000,000
mega	M	10^{6} = 1,000,000
kilo	k	10^{3} = 1,000
hecto	h	10^{2} = 100
deca	da	10^{1} = 10
(unit)		10^{0} = 1

Prefix	Symbol	Multiplication Factor
deci	d	10^{-1} = 0.1
centi	c	10^{-2} = 0.01
milli	m	10^{-3} = 0.001
micro	μ	10^{-6} = 0.000001
nano	n	10^{-9} = 0.000000001
pico	p	10^{-12} = 0.000000000001
femto	f	10^{-15} = 0.000000000000001
atto	a	10^{-18} = 0.000000000000000001

QUESTION POOL SYLLABUS

The syllabus used by the NCVEC Question Pool Committee to develop the question pool is included here as an aid in studying the subelements and topic groups. Reviewing the syllabus will give you an understanding of how the question pool is used to develop the Element 2 written theory examination. Remember, one question will be taken from each topic group within each subelement to create your exam.

Element 2 (Technician Class) Syllabus

T1 – FCC Rules, descriptions and definitions for the amateur radio service, operator and station license responsibilities [6 exam questions - 6 groups]
T1A Amateur Radio services; purpose of the amateur service, amateur-satellite service, operator/primary station license grant, where FCC rules are codified, basis and purpose of FCC rules, meanings of basic terms used in FCC rules
T1B Authorized frequencies; frequency allocations, ITU regions, emission type, restricted sub-bands, spectrum sharing, transmissions near band edges
T1C Operator classes and station call signs; operator classes, sequential, special event, and vanity call sign systems, international communications, reciprocal operation, station license licensee, places where the amateur service is regulated by the FCC, name and address on ULS, license term, renewal, grace period
T1D Authorized and prohibited transmissions
T1E Control operator and control types; control operator required, eligibility, designation of control operator, privileges and duties, control point, local, automatic and remote control, location of control operator
T1F Station identification and operation standards; special operations for repeaters and auxiliary stations, third party communications, club stations, station security, FCC inspection

T2 – Operating Procedures
[3 exam questions - 3 groups]
T2A Station operation; choosing an operating frequency, calling another station, test transmissions, use of minimum power, frequency use, band plans
T2B VHF/UHF operating practices; SSB phone, FM repeater, simplex, frequency offsets, splits and shifts, CTCSS, DTMF, tone squelch, carrier squelch, phonetics
T2C Public service; emergency and non-emergency operations, message traffic handling

T3 – Radio wave characteristics, radio and electromagnetic properties, propagation modes
[3 exam questions - 3 groups]
T3A Radio wave characteristics; how a radio signal travels; distinctions of HF, VHF and UHF; fading, multipath; wavelength vs. penetration; antenna orientation
T3B Radio and electromagnetic wave properties; the electromagnetic spectrum, wavelength vs. frequency, velocity of electromagnetic waves
T3C Propagation modes; line of sight, sporadic E, meteor, aurora scatter, tropospheric ducting, F layer skip, radio horizon

T4 – Amateur radio practices and station setup
[2 exam questions - 2 groups]
T4A Station setup; microphone, speaker, headphones, filters, power source, connecting a computer, RF grounding
T4B Operating controls; tuning, use of filters, squelch, AGC, repeater offset, memory channels

T5 – Electrical principles, math for electronics, electronic principles, Ohm's Law
[4 exam questions - 4 groups]
T5A Electrical principles; current and voltage, conductors and insulators, alternating and direct current
T5B Math for electronics; decibels, electronic units and the metric system
T5C Electronic principles; capacitance, inductance, current flow in circuits, alternating current, definition of RF, power calculations
T5D Ohm's Law

T6 – Electrical components, semiconductors, circuit diagrams, component functions
[4 exam groups - 4 questions]
T6A Electrical components; fixed and variable resistors, capacitors, and inductors; fuses, switches, batteries
T6B Semiconductors; basic principles of diodes and transistors
T6C Circuit diagrams; schematic symbols
T6D Component functions

T7 – Station equipment, common transmitter and receiver problems, antenna measurements and troubleshooting, basic repair and testing
[4 exam questions - 4 groups]
T7A Station radios; receivers, transmitters, transceivers
T7B Common transmitter and receiver problems; symptoms of overload and overdrive, distortion, interference, over and under modulation, RF feedback, off frequency signals; fading and noise; problems with digital communications interfaces
T7C Antenna measurements and troubleshooting; measuring SWR, dummy loads, feedline failure modes
T7D Basic repair and testing; soldering, use of a voltmeter, ammeter, and ohmmeter

T8 – Modulation modes, amateur satellite operation, operating activities, non-voice communications
[4 exam questions - 4 groups]
T8A Modulation modes; bandwidth of various signals
T8B Amateur satellite operation; Doppler shift, basic orbits, operating protocols
T8C Operating activities; radio direction finding, radio control, contests, special event stations, basic linking over Internet
T8D Non-voice communications; image data, digital modes, CW, packet, PSK31

T9 – Antennas, feed lines
[2 exam groups - 2 questions]
T9A Antennas; vertical and horizontal, concept of gain, common portable and mobile antennas, relationships between antenna length and frequency
T9B Feed lines; types, losses vs. frequency, SWR concepts, matching, weather protection, connectors

T0 – AC power circuits, antenna installation, RF hazards [3 exam questions - 3 groups]
T0A AC power circuits; hazardous voltages, fuses and circuit breakers, grounding, lightning protection, battery safety, electrical code compliance
T0B Antenna installation; tower safety, overhead power lines
T0C RF hazards; radiation exposure, proximity to antennas, recognized safe power levels, exposure to others

2010-14 ELEMENT 2 Q&A CROSS REFERENCE

The following cross reference presents all 394 question numbers in numerical order included in the 2010-14 Element 2 Question Pool, followed by the page number on which the question begins in the book. This will allow you to locate specific questions by question number.

T1 – FCC Rules, descriptions and definitions for the amateur radio service, operator and station license responsibilities

Question	Page
T1A01	31
T1A02	31
T1A03	49
T1A04	51
T1A05	99
T1A06	104
T1A07	103
T1A08	84
T1A09	84
T1A10	34
T1A11	85
T1B01	39
T1B02	40
T1B03	58
T1B04	59
T1B05	60
T1B06	61
T1B07	60
T1B08	62
T1B09	62
T1B10	59
T1B11	59
T1C01	36
T1C02	35
T1C03	40
T1C04	42
T1C05	62
T1C06	41
T1C07	53
T1C08	32
T1C09	33
T1C10	32
T1C11	33
T1D01	40
T1D02	43
T1D03	51
T1D04	52
T1D05	52
T1D06	49
T1D07	80
T1D08	47
T1D09	50
T1D10	50
T1D11	51
T1E01	44
T1E02	44
T1E03	45
T1E04	45
T1E05	46
T1E06	46
T1E07	45
T1E08	47
T1E09	46
T1E10	47
T1E11	47
T1F01	38
T1F02	38
T1F03	35
T1F04	38
T1F05	85
T1F06	39
T1F07	39
T1F08	46
T1F09	79
T1F10	48
T1F11	40
T1F12	37
T1F13	53

T2 – Operating Procedures

Question	Page
T2A01	81
T2A02	60
T2A03	81
T2A04	73
T2A05	73
T2A06	71
T2A07	71
T2A08	71
T2A09	83
T2A10	61
T2A11	50
T2B01	70
T2B02	82
T2B03	70
T2B04	81
T2B05	118
T2B06	118
T2B07	120
T2B08	74
T2B09	36
T2B10	74
T2B11	74
T2C01	88
T2C02 *deleted by QPC*	
T2C03 *deleted by QPC*	
T2C04	89
T2C05	88
T2C06	86
T2C07	88
T2C08	90
T2C09	87
T2C10	89
T2C11	90

T3 – Radio wave characteristics, radio and electromagnetic properties, propagation modes

Question	Page
T3A01	73
T3A02	78
T3A03	156
T3A04	157
T3A05	156
T3A06	73
T3A07	55
T3A08	98
T3A09	97
T3A10	107
T3A11	96
T3B01	56
T3B02	128
T3B03	55
T3B04	55
T3B05	57
T3B06	57
T3B07	56
T3B08	58
T3B09	60
T3B10	58
T3B11	55
T3C01	98
T3C02	96
T3C03	95
T3C04	97
T3C05	94
T3C06	94
T3C07	95
T3C08	95
T3C09	96
T3C10	93
T3C11	94

T4 – Amateur radio practices and station setup

Question	Page
T4A01	65
T4A02	65
T4A03	65
T4A04	122
T4A05	122
T4A06	106
T4A07	106
T4A08	172
T4A09	121
T4A10	121
T4A11	127
T4B01	120
T4B02	69
T4B03	69
T4B04	63
T4B05	121
T4B06	116
T4B07	117
T4B08	116
T4B09	116
T4B10	113
T4B11	80

T5 – Electrical principles, math for electronics, electronic principles, Ohm's Law

Question	Page
T5A01	128
T5A02	137
T5A03	127
T5A04	129
T5A05	126
T5A06	127
T5A07	128
T5A08	130
T5A09	128
T5A10	137
T5A11	126

Glossary

Amateur communication: Noncommercial radio communication by or among amateur stations solely with a personal aim and without personal or business interest.

Amateur operator/primary station license: An instrument of authorization issued by the FCC comprised of a station license, and also incorporating an operator license indicating the class of privileges.

Amateur operator: A person holding a valid license to operate an amateur station issued by the FCC. Amateur operators are frequently referred to as ham operators.

Amateur Radio services: The amateur service, the amateur-satellite service, and the radio amateur civil emergency service.

Amateur-satellite service: A radiocommunication service using stations on Earth satellites for the same purpose as those of the amateur service.

Amateur service: A radiocommunication service for the purpose of self-training, intercommunication and technical investigations carried out by amateurs; that is, duly authorized persons interested in radio technique solely with a personal aim and without pecuniary interest.

Amateur station: A station licensed in the amateur service embracing necessary apparatus at a particular location used for amateur communication.

AMSAT: Radio Amateur Satellite Corporation, a nonprofit scientific organization. (P.O. Box #27, Washington, DC 20044)

ANSI: American National Standards Institute. A non-government organization that develops recommended standards for a variety of applications.

APRS: Automatic Position Radio System, which takes GPS (Global Positioning System) information and translates it into an automatic packet of digital information.

ARES: Amateur Radio Emergency Service — the emergency division of the American Radio Relay League. Also see RACES

ARRL: American Radio Relay League, national organization of U.S. Amateur Radio operators. (225 Main Street, Newington, CT 06111)

Audio Frequency (AF): The range of frequencies that can be heard by the human ear, generally 20 hertz to 20 kilohertz.

Automatic control: The use of devices and procedures for station control without the control operator being present at the control point when the station is transmitting.

Automatic Volume Control (AVC): A circuit that continually maintains a constant audio output volume in spite of deviations in input signal strength.

Beam or Yagi antenna: An antenna array that receives or transmits RF energy in a particular direction. Usually rotatable.

Block diagram: A simplified outline of an electronic system where circuits or components are shown as boxes.

Broadcasting: Information or programming transmitted by radio means intended for the general public.

Bulletin No. 65: The Office of Engineering & Technology bulletin that provides specified safety guidelines for human exposure to radiofrequency (RF) radiation.

Business communications: Any transmission or communication the purpose of which is to facilitate the regular business or commercial affairs of any party. Business communications are prohibited in the amateur service.

Call Book: A published list of all licensed amateur operators available in North American and Foreign editions.

Call sign: The FCC systematically assigns each amateur station its primary call sign.

Certificate of Successful Completion of Examination (CSCE): A certificate providing examination credit for 365 days. Both written and code credit can be authorized.

Coaxial cable, Coax: A concentric, two-conductor cable in which one conductor surrounds the other, separated by an insulator.

Controlled Environment: Involves people who are aware of and who can exercise control over radiofrequency exposure. Controlled exposure limits apply to both occupational workers and Amateur Radio operators and their immediate households.

Control operator: An amateur operator designated by the licensee of an amateur station to be responsible for the station transmissions.

Coordinated repeater station: An amateur repeater station for which the transmitting and receiving frequencies have been recommended by the recognized repeater coordinator.

Coordinated Universal Time (UTC): (Also Greenwich Mean Time, UCT or Zulu time.) The time at the zero-degree (0°) Meridian which passes through Greenwich, England. A universal time among all amateur operators.

Crystal: A quartz or similar material which has been ground to produce natural vibrations of a specific frequency. Quartz crystals produce a high degree of frequency stability in radio transmitters.

CW: See Morse code.

Dipole antenna: The most common wire antenna. Length is equal to one-half of the wavelength. Fed by coaxial cable.

Dummy antenna: A device or resistor which serves as a transmitter's antenna without radiating radio waves. Generally used to tune up a radio transmitter.

Duplexer: A device that allows a single antenna to be simultaneously used for both reception and transmission.

Duty cycle: As applies to RF safety, the percentage of time that a transmitter is "on" versus "off" in a 6- or 30-minute time period.

Effective Radiated Power (ERP): The product of the transmitter (peak envelope) power, expressed in watts, delivered to the antenna, and the relative gain of an antenna over that of a half-wave dipole antenna.

Electromagnetic radiation: The propagation of radiant energy, including infrared, visible light, ultraviolet, radiofrequency, gamma and X-rays, through space and matter.

Emergency communication: Any amateur communication directly relating to the immediate safety of life of individuals or the immediate protection of property.

Examination Element: The written theory exam or CW test required for various classes of FCC Amateur Radio licenses. Technician must pass Element 2 written theory; General must pass Element 3 written theory plus Element 1 CW; Extra must pass Element 4 written theory.

Far Field: The electromagnetic field located at a great distance from a transmitting antenna. The far field begins at a distance that depends on many factors, including the wavelength and the size of the antenna. Radio signals are normally received in the far field.

FCC Form 605: The FCC application form used to apply for a new amateur operator/primary station license or to renew or modify an existing license.

Federal Communications Commission (FCC): A board of five Commissioners, appointed by the President, having the power to regulate wire and radio telecommunications in the U.S.

Feedline: A system of conductors that connects an antenna to a receiver or transmitter.

Field Day: Annual activity sponsored by the ARRL to demonstrate emergency preparedness of amateur operators.

Field strength: A measure of the intensity of an electric or magnetic field. Electric fields are measured in volts per meter; magnetic fields in amperes per meter.

Filter: A device used to block or reduce alternating currents or signals at certain frequencies while allowing others to pass unimpeded.

Frequency: The number of cycles of alternating current in one second.

Frequency coordinator: An individual or organization which recommends frequencies and other operating and/or technical parameters for amateur repeater operation in order to avoid or minimize potential interferences.

Frequency Modulation (FM): A method of varying a radio carrier wave by causing its frequency to vary in accordance with the information to be conveyed.

Frequency privileges: The transmitting frequency bands available to the various classes of amateur operators. The various Class privileges are listed in Part 97.301 of the FCC rules.

Ground: A connection, accidental or intentional, between a device or circuit and the earth or some common body and the earth or some common body serving as the earth.

Ground wave: A radio wave that is propagated near or at the earth's surface.

Handi-Ham system: Amateur organization dedicated to assisting handicapped amateur operators. (3915 Golden Valley Road, Golden Valley, MN 55422)

Harmful interference: Interference which seriously degrades, obstructs or repeatedly interrupts the operation of a radio communication service.

Harmonic: A radio wave that is a multiple of the fundamental frequency. The second harmonic is twice the fundamental frequency, the third harmonic, three times, etc.

Hertz: One complete alternating cycle per second. Named after Heinrich R. Hertz, a German physicist. The number of hertz is the frequency of the audio or radio wave.

High Frequency (HF): The band of frequencies that lie between 3 and 30 Megahertz. It is from these frequencies that radio waves are returned to earth from the ionosphere.

High-Pass filter: A device that allows passage of high frequency signals but attenuates the lower frequencies. When installed on a television set, a high-pass filter allows TV frequencies to pass while blocking lower-frequency amateur signals.

Inverse Square Law: The physical principle by which power density decreases as you get further away from a transmitting antenna. RF power density decreases by the inverse square of the distance.

Ionization: The process of adding or stripping away electrons from atoms or molecules. Ionization occurs when substances are heated at high temperatures or exposed to high voltages. It can lead to significant genetic damage in biological tissue.

Ionosphere: Outer limits of atmosphere from which HF amateur communications signals are returned to earth.

IRC: International Reply Coupon, a method of prepaying postage for a foreign amateur's QSL card.

Jamming: The intentional, malicious interference with another radio signal.

Key clicks, Chirps: Defective keying of a telegraphy signal sounding like tapping or high varying pitches.

Linear amplifier: A device that accurately reproduces a radio wave in magnified form.

Long wire: A horizontal wire antenna that is one wavelength or longer in length.

Low-Pass filter: Device connected to worldwide transmitters that inhibits passage of higher frequencies that cause television interference but does not affect amateur transmissions.

Machine: A ham slang word for an automatic repeater station.

Malicious interference: See jamming.

MARS: The Military Affiliate Radio System. An organization that coordinates the activities of amateur communications with military radio communications.

Maximum authorized transmitting power: Amateur stations must use no more than the maximum transmitter power necessary to carry out the desired communications. The maximum P.E.P. output power levels authorized Novices are 200 watts in the 80-, 40-, 15- and 10-meter bands, 25 watts in the 222-MHz band, and 5 watts in the 1270-MHz bands.

Maximum Permissible Exposure (MPE): The maximum amount of electric and magnetic RF energy to which a person may safely be exposed.

Maximum usable frequency (MFU): The highest frequency that will be returned to earth from the ionosphere.

Medium frequency (MF): The band of frequencies that lies between 300 and 3,000 kHz (3 MHz).

Microwave: Electromagnetic waves with a frequency of 300 MHz to 300 GHz. Microwaves can cause heating of biological tissue.

Mobile operation: Radio communications conducted while in motion or during halts at unspecified locations.

Mode: Type of transmission such as voice, teletype, code, television, facsimile.

Modulate: To vary the amplitude, frequency, or phase of a radiofrequency wave in accordance with the information to be conveyed.

Morse code: The International Morse code, A1A emission. Interrupted continuous wave communications conducted using a dot-dash code for letters, numbers and operating procedure signs.

Near Field: The electromagnetic field located in the immediate vicinity of the antenna. Energy in the near field depends on the size of the antenna, its wavelength and transmission power.

Nonionizing radiation: Electromagnetic waves, or fields, which do not have the capability to alter the molecular structure of substances. RF energy is nonionizing radiation.

Novice operator: An FCC licensed, entry-level amateur operator in the amateur service.

Occupational exposure: See controlled environment.

OET: Office of Engineering & Technology, a branch of the FCC that has developed the guidelines for radiofrequency (RF) safety.

Ohm's law: The basic electrical law explaining the relationship between voltage, current and resistance. The current (I) in a circuit is equal to the voltage (E) divided by the resistance (R), or $I = E/R$.

OSCAR: "Orbiting Satellite Carrying Amateur Radio." A series of satellites designed and built by amateur operators of several nations.

Oscillator: A device for generating oscillations or vibrations of an audio or radiofrequency signal.

Packet radio: A digital method of communicating computer-to-computer. A terminal-node controller makes up the packet of data and directs it to another packet station.

Peak Envelope Power (PEP): 1. The power during one radiofrequency cycle at the crest of the modulation envelope, taken under normal operating conditions. 2. The maximum power that can be obtained from a transmitter.

Phone patch: Interconnection of amateur radio to the public switched telephone network, and operated by the control operator of the station.

Power density: A measure of the strength of an electromagnetic field at a distance from its source. Usually expressed in milliwatts per square centimeter (mW/cm2). Far-field power density decreases according to the Law of Inverse Squares.

Power supply: A device or circuit that provides the appropriate voltage and current to another device or circuit.

Propagation: The travel of electromagnetic waves or sound waves through a medium.

Public exposure: See "uncontrolled" environment.

Q-signals: International three-letter abbreviations beginning with the letter Q used primarily to convey information using the Morse code.

QSL Bureau: An office that bulk processes QSL (radio confirmation) cards for (or from) foreign amateur operators as a postage-saving mechanism.

RACES (Radio Amateur Civil Emergency Service): A radio service using amateur stations for civil defense communications during periods of local, regional, or national emergencies.

Radiation: Electromagnetic energy, such as radio waves, traveling forth into space from a transmitter.

Radiofrequency (RF): The range of frequencies over 20 kilohertz that can be propagated through space.

Radiofrequency (RF) radiation: Electromagnetic fields or waves having a frequency between 3 kHz and 300 GHz.

Radiofrequency spectrum: The eight electromagnetic bands ranked according to their frequency and wavelength. Specifically, the very-low, low, medium, high, very-high, ultra-high, super-high, and extremely-high frequency bands.

Radio wave: A combination of electric and magnetic fields varying at a radiofrequency and traveling through space at the speed of light.

Repeater operation: Automatic amateur stations that retransmit the signals of other amateur stations.

Routine RF radiation evaluation: The process of determining if the RF energy from a transmitter exceeds the Maximum Permissible Exposure (MPE) limits in a controlled or uncontrolled environment.

RST Report: A telegraphy signal report system of Readability, Strength and Tone.

S-meter: A voltmeter calibrated from 0 to 9 that indicates the relative signal strength of an incoming signal at a radio receiver.

Selectivity: The ability of a circuit (or radio receiver) to separate the desired signal from those not wanted.

Sensitivity: The ability of a circuit (or radio receiver) to detect a specified input signal.

Short circuit: An unintended, low-resistance connection across a voltage source resulting in high current and possible damage.

Shortwave: The high frequencies that lie between 3 and 30 Megahertz that are propagated long distances.

Single-Sideband (SSB): A method of radio transmission in which the RF carrier and one of the sidebands is suppressed and all of the information is carried in the one remaining sideband.

Skip wave, Skip zone: A radio wave reflected back to earth. The distance between the radio transmitter and the site of a radio wave's return to earth.

Sky-wave: A radio wave that is refracted back to earth. Sometimes called an ionospheric wave.

Specific Absorption Rate (SAR): The time rate at which radiofrequency energy is absorbed into the human body.

Spectrum: A series of radiated energies arranged in order of wavelength. The radio spectrum extends from 20 kilohertz upward.

Spurious Emissions: Unwanted radiofrequency signals emitted from a transmitter that sometimes cause interference.

Station license, location: No transmitting station shall be operated in the amateur service without being licensed by the FCC. Each amateur station shall have one land location, the address of which appears in the station license.

Sunspot Cycle: An 11-year cycle of solar disturbances which greatly affects radio wave propagation.

Technician operator: An Amateur Radio operator who has successfully passed Element 2.

Technician-Plus: An amateur operator who has passed a 5-wpm code test in addition to Technician Class requirements.

Telegraphy: Communications transmission and reception using CW, International Morse code.

Telephony: Communications transmission and reception in the voice mode.

Telecommunications: The electrical conversion, switching, transmission and control of audio video and data signals by wire or radio.

Temporary operating authority: Authority to operate your amateur station while awaiting arrival of an upgraded license.

Terrestrial station location: Any location of a radio station on the surface of the earth including the sea.

Thermal effects: As applies to RF radiation, biological tissue damage resulting because of the body's inability to cope with or dissipate excessive heat.

Third-party traffic: Amateur communication by or under the supervision of the control operator at an amateur station to another amateur station on behalf of others.

Time-averaging: As applies to RF safety, the amount of electromagnetic radiation over a given time. The premise of time-averaging is that the human body can tolerate the thermal load caused by high, localized RF exposures for short periods of time.

Transceiver: A combination radio transmitter and receiver.

Transition region: Area where power density decreases inversely with distance from the antenna.

Transmatch: An antenna tuner used to match the impedance of the transmitter output to the transmission line of an antenna.

Transmitter: Equipment used to generate radio waves. Most commonly, this radio carrier signal is amplitude varied or frequency varied (modulated) with information and radiated into space.

Transmitter power: The average peak envelope power (output) present at the antenna terminals of the transmitter. The term "transmitted" includes any external radio-frequency power amplifier which may be used.

Ultra High Frequency (UHF): Ultra high frequency radio waves that are in the range of 300 to 3,000 MHz.

Uncontrolled environment: Applies to those persons who have no control over their exposure to RF energy in the environment. Residences adjacent to ham radio installations are considered to be in an "uncontrolled" environment.

Upper Sideband (USB): The proper operating mode for sideband transmissions made in the new Novice 10-meter voice band. Amateurs generally operate USB at 20 meters and higher frequencies; lower sideband (LSB) at 40 meters and lower frequencies.

Very High Frequency (VHF): Very high frequency radio waves that are in the range of 30 to 300 MHz.

Volunteer Examiner: An amateur operator of at least a General Class level who prepares and administers amateur operator license examinations.

Volunteer Examiner Coordinator (VEC): A member of an organization which has entered into an agreement with the FCC to coordinate the efforts of volunteer examiners in preparing and administering examinations for amateur operator licenses.

Index

A special invitation from Gordon West, WB6NOA, ARRL Life Member

"Join ARRL and experience the BEST of Ham Radio!"

FREE Book Offer!

ARRL Membership Benefits and Services:

- *QST* magazine — your monthly source of news, easy-to-read product reviews, and features for new hams!
- Technical Information Service — access to problem-solving experts!
- Members-only Web services — find information fast, anytime!
- ARRL clubs, mentors and volunteers — ready to welcome YOU!

I want to join ARRL.

Send me the FREE book I have selected
(choose one)

☐ **The ARRL Repeater Directory**
☐ **Getting Started with Ham Radio**
☐ **The ARRL Emergency Communication Handbook**

Name ______________________ **Call Sign** ______________

Street ____

City ______________________ **State** __________ **ZIP** __________

Email ____

Please check the appropriate one-year[1] rate:

❑ **$39 in US.**
❑ **Age 21 or younger rate, $20 in US** (see note*).
❑ **Canada $49.**
❑ **Elsewhere $62.**

[1] 1-year membership dues include $15 for a 1-year subscription to QST. International 1-year rates include a $10 surcharge for surface delivery to Canada and a $23 surcharge for air delivery to other countries. Other US membership options available: Blind, Life, and QST by First Class postage. Contact ARRL for details.
*Age 21 or younger rate applies only if you are the oldest licensed amateur in your household.
International membership is available with an annual CD-ROM option (no monthly receipt of QST). Contact ARRL for details.
Dues subject to change without notice.

Sign up my family members, residing at the same address, as ARRL members too! They'll each pay only $8 for a year's membership, have access to ARRL benefits and services (except QST) and also receive a membership card.

❑ Sign up ______ family members @ $8 each = $ ________. Attach their names & call signs (if any).

❑ Total amount enclosed, payable to ARRL $ _______________ . (US funds drawn on a bank in the US).

❑ Enclosed is $ ___________ ($1.00 minimum) as a donation to the Legal Research and Resource Fund.

❑ Charge to: ❑ VISA ❑ MasterCard ❑ Amex ❑ Discover

Card Number ______________________ Expiration Date __________

Cardholder's Signature ______________________

Call Toll-Free (US) **1-888-277-5289**
Join Online **www.arrl.org/join** or
Clip and send to:

ARRL *The national association for AMATEUR RADIO™*
225 Main Street
Newington, CT 06111-1494 USA

❑ If you do not want your name and address made available for non-ARRL related mailings, please check here.

Web Code: GW

$50.00
OFF ANY BASE*

$20.00
OFF ANY MOBILE*
(Includes IC-706MKIIG)

Gordo Bucks

$10.00
OFF ANY HANDHELD*

INSTRUCTIONS:
*This coupon is good for new Icom amateur and receiver products only. Offer available to students of Gordon West's amateur radio technician class. Completion of class must be earned no earlier then 90 days before date of purchase. Valid at any participating authorized Icom dealer. Limit 1 coupon per person.

Icom Dealer: please attach a copy of the Gordon West class completion certificate along with a copy of the invoice with this coupon when submitting for reimbursement from Icom America Inc.

See page 192 on how to obtain your personally signed Gordon West Technician Class Completion Certificate necessary when you present this generous discount coupon from Icom America.

TAPE

FREE
CQ Amateur Radio Mini-Sub!

Fill in your address information on the reverse side of this page, fold in half, tape closed and mail today!

CUT HERE

TAPE

TAPE

NO POSTAGE NECESSARY IF MAILED IN THE UNITED STATES

BUSINESS REPLY MAIL

FIRST CLASS MAIL PERMIT NO. 2055 HICKSVILLE, NY 11801

POSTAGE WILL BE PAID BY ADDRESSEE

CQ **The Radio Amateur's Journal**

25 Newbridge Road
Hicksville, NY 11801-9962